Large-Scale Landscape Experiments

Landscape change and habitat fragmentation are key factors impacting biodiversity worldwide. These processes have many facets, each of which is usually studied in isolation. The Tumut Fragmentation Study has run for 13 years and has yielded extensive data on changes in both plant and animal populations in areas of native forest and pine plantation. It is unique in the way that many different factors and processes influencing a wide range of species groups have been studied in a single large-scale natural experiment. Writing for academic researchers, professionals and graduate students, David B. Lindenmayer uses the Tumut Fragmentation Study and other relevant research to provide an overview of the relationships between landscape change, habitat fragmentation, and biodiversity conservation. He details the background ecology of landscape change and habitat fragmentation, the experimental design underpinning the establishment of the large-scale natural experiment, the implementation of, and results from, an array of key and strongly contrasting research programmes over the past 13 years. Key lessons are drawn on throughout the book on the design and implementation of large-scale ecological studies, biodiversity conservation in fragmented landscapes and the management of plantation landscapes for enhanced nature conservation. The book highlights how important new insights can be generated from integrating demographic, genetic and modelling research.

DAVID B. LINDENMAYER is Professor of Conservation Science and Ecology in the Fenner School of Environment and Society at The Australian National University, Canberra.

ECOLOGY, BIODIVERSITY AND CONSERVATION

The world's biological diversity faces unprecedented threats. The urgent challenge facing the concerned biologist is to understand ecological processes well enough to maintain their functioning in the face of the pressures resulting from human population growth. Those concerned with the conservation of biodiversity and with restoration also need to be acquainted with the political, social, historical, economic and legal frameworks within which ecological and conservation practice must be developed. The new Ecology, Biodiversity and Conservation series will present balanced, comprehensive, up-to-date, and critical reviews of selected topics within the sciences of ecology and conservation biology, both botanical and zoological, and both 'pure' and 'applied'. It is aimed at advanced final-year undergraduates, graduate students, researchers and university teachers, as well as ecologists and conservationists in industry, government and the voluntary sectors. The series encompasses a wide range of approaches and scales (spatial, temporal, and taxonomic), including quantitative, theoretical, population, community, ecosystem, landscape, historical, experimental, behavioural and evolutionary studies. The emphasis is on science related to the real world of plants and animals rather than on purely theoretical abstractions and mathematical models. Books in this series will, wherever possible, consider issues from a broad perspective. Some books will challenge existing paradigms and present new ecological concepts, empirical or theoretical models, and testable hypotheses. Other books will explore new approaches and present syntheses on topics of ecological importance.

The Ecology of Phytoplankton
C.S. Reynolds

Invertebrate Conservation and Agricultural Ecosystems
T.R. New

Risks and Decisions for Conservation and Environmental Management
Mark Burgman

Large-Scale Landscape Experiments

Experiments

Lessons from Tumut

DAVID B. LINDENMAYER
The Australian National University

CAMBRIDGE
UNIVERSITY PRESS

University Printing House, Cambridge CB2 8BS, United Kingdom

One Liberty Plaza, 20th Floor, New York, NY 10006, USA

477 Williamstown Road, Port Melbourne, VIC 3207, Australia

4843/24, 2nd Floor, Ansari Road, Daryaganj, Delhi - 110002, India

79 Anson Road, #06-04/06, Singapore 079906

Cambridge University Press is part of the University of Cambridge.

It furthers the University's mission by disseminating knowledge in the pursuit of education, learning and research at the highest international levels of excellence.

www.cambridge.org
Information on this title: www.cambridge.org/9780521707787

First published 2009

A catalogue record for this publication is available from the British Library

Library of Congress Cataloging in Publication data
Lindenmayer, David,
Seeing the forest and the trees : lessons on forest fragmentation, ecology, conservation, and plantation management from the Tumut fragmentation study / David B. Lindenmayer
    p.   cm.
Includes bibliographical references and index.
1. Fragmented landscapes – Australia – Tumut (N.S.W.) Region.   2. Forest ecology – Australia – Tumut (N.S.W.) Region.   3. Forest management – Environmental aspects – Australia – Tumut (N.S.W.) Region.
I. Title
QH197.2.N48L56   2009
577.27–dc22
2008053642

ISBN   978-0-521-88156-2   Hardback
ISBN   978-0-521-70778-7   Paperback

For Ross Cunningham
whose statistical insights and mentoring turned a good idea into reality.
For Karen Viggers
with thanks for her great patience with book-writing projects over the years.

Contents

Preface

The Tumut Fragmentation Study is a relatively rare beast in the world of ecological research – it's a large-scale, long-term ecological research programme that incorporates observational studies that have run for over a decade, experimental manipulations and integrated demographic and genetic research. The research programme has provided a platform for a range of simulation modelling studies. The Tumut Fragmentation Study has also served as a wonderful test bed for a broad range of ecological theory and a platform for modelling. And it all began with a glance out of a plane window.

In the early 1990s, I was flying to Melbourne from Canberra when I saw a fascinating landscape near the timber town of Tumut in southern New South Wales, south-eastern Australia. I saw vast areas of land in which the native forest had been cleared for pine plantation, and set in this plantation were numerous pockets of native forest that were now 'islands in a sea of pines'. The exciting thing was that these 'islands' were not the unwanted parts of the landscape that you often find in pine plantations where, for example, native remnants are only on rocky hills. Nor were these remnants only representing riparian areas. They were, in fact, patches of forest of the same type as the nearby intact native forest.

And from this observation immediately sprang a series of questions: How does the biodiversity of these native remnants compare with the biodiversity of the intact native forest? And, flowing on from this, what contribution might these remnants make to conserving biodiversity in production landscapes such as plantations? What role might they play in assisting animal dispersal and movement in fragmented landscapes?

And so the idea for the Tumut Fragmentation Study was born. It officially commenced in 1995.

The study is focused on the Buccleuch State Forest 100 km west of Canberra. The forest is a 50,000 ha plantation of Radiata Pine established by clearing native forest from 1932 until 1985. In the process, however, patches of native bush ranging in size from half a hectare to 125 ha were left in place. And beyond the boundaries of the plantation are large continuous areas of native eucalypt forest.

Field sites were established in the Radiata Pine stands (the plantation), in large unfragmented forest (the control sites) and in the native forest 'islands' within pine stands (treatment sites). Initially data were gathered on the presence and abundance of arboreal marsupials, small mammals, forest birds and plants at each of the sites. However, the study grew to include a range of other studies that examined aspects ranging from how different groups of animals behaved in and around the patches to what was the genetic make-up of the different organisms using these remnants.

The loss, fragmentation and degradation of native vegetation is the single greatest driver of biodiversity decline in Australia. Finding ways to improve the quantity and quality of native vegetation with optimal conservation outcomes is critical to conserving biodiversity. The Tumut Fragmentation Study is a rare long-term, large-scale ecological investigation that provides valuable empirical evidence on the ecological processes at play and how we might best manage our production landscapes to ameliorate the loss of biodiversity.

The data generated by the Tumut Fragmentation Study have led to the publication of over a hundred journal papers and have contributed to four books. However, while the study is still a 'work in progress', there has been a growing call to write a book on what has been achieved, in order to draw together the various strands of work and outline some of lessons learned. This volume is an attempt to do this.

The book that has resulted isn't perfect. Some will want more detail, while others will find it overly technical. However, for anyone interested in the science of landscape change, biodiversity conservation and the realities of large-scale ecological studies, this overview of the Tumut Fragmentation Study will hopefully be an illuminating read.

It's not often that we have the opportunity to explore landscape change over scales of hundreds of kilometres and time frames of decades. It's my hope that the lessons presented here, gleaned from many years of hard work by many researchers, will benefit us all through the better conservation of our precious biodiversity.

David Lindenmayer

Acknowledgements

The Tumut project owes its origins, conceptual development and much of its statistical design to Adjunct Professor Ross Cunningham. His insights and mentorship helped convert what seemed like good ideas on paper to a reality on the ground.

In 2006, a close friend and colleague, Professor Jerry Franklin, insisted that it was time to draw together the published material and general insights from Tumut into one place. That idea was then supported by the Joint Venture Agroforestry Program (JVAP) within the Rural Industries Research and Development Corporation, based in Canberra, Australia. In particular, Dr Rosemary Lott, from that organisation, championed this writing project.

A vast number of people and organisations have made major contributions to the research at Tumut (see Appendix 1). In particular, field staff have been pivotal to the success of the project, especially Mason Crane, Chris MacGregor, Damian Michael, Dr. Rebecca Montague-Drake, Matthew Pope, David Rawlins and Craig Tribolet.

A great deal of excellent work has been completed at Tumut by postgraduate students. They include: Dr Sam Banks, Dr Joern Fischer, Ms Heidi Hewittson, Mr Matthew Pope, Dr Monica Ruibal, Dr Peter Smith, Dr Jeanette Stanley, Dr Darius Tubelis and Ms Kara Yongentob.

Work at Tumut has progressed through collaborative partnerships with many colleagues (see Appendix 1). These include: Dr Sam Banks, Dr Andrew Claridge, Adjunct Professor Ross Cunningham, Ms Christine Donnelly, Dr Joern Fischer, Dr Phil Gibbons, Professor Richard Hobbs, Dr Bob Lacy, Dr Sarah Legge, Professor Mike McCarthy, Dr Sue McIntyre, Dr Kirtsen Parris, Dr Rod Peakall, Dr Emma Pharo, Professor Hugh Possingham, Dr Peter Smith, Dr Paul Sunnucks, Dr Andrea Taylor, Professor Hugh Tyndale-Biscoe, Dr Karen Viggers, Professor Alan Welsh and Dr Jeff Wood.

Studies at Tumut have been supported by grants from many bodies. These include: Land and Water Australia, the Joint Venture Agroforestry

Program (JVAP), The Australian Research Council, the former NSW Department of Conservation and Land Management, Department of Environment and Water Resources, the Pratt Foundation, State Forests of New South Wales, the former NSW National Parks and Wildlife Service, the Winifred Violet Scott Trust, the Kendall Foundation and the now defunct CSR Ltd.

Volunteers have made an enormous contribution to the work at Tumut. In particular the efforts of the dedicated birdwatchers from the Canberra Ornithologists Group (COG) have been invaluable. My father, Bruce Lindenmayer, has been instrumental in gathering support from many outstanding amateur ornithologists.

David Salt edited the book and made many valuable changes that greatly improved a manuscript that started out much sloppier than it should have been! This book could not have been produced without the heroic efforts of Rachel Muntz who contributed her considerable expertise to many facets of this work.

Finally, Alan Crowden steered this book through the process of conception, contract development and those other fiddly stages that are always more convoluted than they initially seem.

1 · The science of understanding landscape change: setting the scene for the Tumut Fragmentation Study

The body of work that comprises the Tumut Fragmentation Study is perhaps different from other projects around the world in that a wide range of questions have been posed and many kinds of work have been undertaken in the same place. The aim of this short book is to draw together some of the key insights from this work and outline some of the lessons learned. The hope is that a synthesis from publications scattered around in different journals and books might be more than the 'sum of its parts'. At its heart, the Tumut Fragmentation Study is all about understanding landscape change and biodiversity.

Landscape change and habitat fragmentation are processes that pose a major threat to many species. Because of this, an enormous number of studies worldwide have investigated how flora and fauna respond to them. Many of these studies have attempted to determine ways in which conservation can be more effective in areas extensively modified by humans. Themes associated with landscape change and habitat fragmentation have therefore become a major focus of conservation biology and landscape ecology (McGarigal and Cushman, 2002; Fahrig, 2003; Hobbs and Yates, 2003) and are now two of the most frequently studied processes threatening species persistence (Fazey *et al.*, 2005; Lindenmayer and Fischer, 2006).

A huge range of studies and an enormous variety of topics fall under the broad umbrella of landscape change and habitat fragmentation. Research can be focused on single-species responses, communities or aggregate species richness. Single-species investigations often highlight the fact that each individual species responds uniquely to landscape change. However, a detailed focus on the response of an individual species may tell us little about the overall pattern of change in larger assemblages of species, and the management reality is that it is rarely possible to consider

---

Box 1.1. What's in a name?

The initial (aerial) perspective of the landscape at Tumut was an impression of patches of remnant native forest surrounded by stands of exotic Radiata Pine plantation. From a human perspective, the original forest cover had been broken into patches – a process often termed 'fragmentation' or 'habitat fragmentation' by conservation biologists. The obvious name for the research programme was the Tumut Fragmentation Study. While the name has stuck, its validity has become problematic as new insights from the work have been gained over more than a decade of research. This is because the landscape is clearly not 'fragmented' for some species which readily live in the pine stands – they may even be more common in pine plantations than in eucalypt forests. Thus, for some species, the eucalypt remnants may not behave as 'fragments' of forest in a 'sea of inhospitable pine stands'. In essence, the human perspective of a 'fragmented landscape' may not be one shared by many elements of the biota. We return to this important insight several times throughout this book because it has major implications for the different kinds of conceptual models that can be employed. Despite these problems concerning the name 'Tumut Fragmentation Study', its use has been retained for the sake of simplicity throughout this book.

---

more than a handful of species in any given area. Thus, the identification of general patterns involving many species can be useful from a management perspective, although this often makes highly simplifying assumptions about species co-occurrence patterns (Lindenmayer and Fischer, 2007).

Research on landscape change and habitat fragmentation can also be focused on either spatial patterns or ecological processes. This is because landscape change can result in altered patterns of native vegetation cover and altered ecological processes. Both processes and patterns can be significantly correlated with various measures of biota. Patterns are directly observable because they are made up of the configuration of one or more entities. By contrast, processes can only be inferred indirectly through the patterns they produce. Nevertheless, a better understanding of the processes giving rise to emergent patterns may help with the understanding of how best to extrapolate findings from one landscape to another that has not previously been studied (Lindenmayer and Fischer, 2007).

Within the continuum of themes of individual species and assemblage-level responses, and patterns and processes, the past decade of research at Tumut has encompassed many different kinds of work. These have included studies of:

- **Individual species**, such as the response of individual mammal or reptile taxa to landscape context effects (Lindenmayer *et al.*, 1999a; Fischer *et al.*, 2005; Chapter 5);
- **Assemblages of species**, e.g. frogs (Parris and Lindenmayer, 2004; Chapter 5);
- **Overall species richness**, e.g. vascular plant species richness (Smith, 2006; Chapter 5);
- **Ecological patterns**, e.g. the nestedness of vertebrate assemblages (Fischer and Lindenmayer, 2005a; Chapter 7);
- **Ecological processes**, e.g. the mechanisms of population recovery following patch-level perturbation (Lindenmayer *et al.*, 2005a; Peakall and Lindenmayer, 2006; Chapter 9).

The work at Tumut has also involved an array of approaches such as:

- **True experiments** in which active manipulations of the landscape or vertebrate populations within patches have taken place (e.g. the Edge Experiment, Lindenmayer *et al.*, 2008a; Chapter 10);
- **Natural experiments** where active manipulation has occurred but was not controlled by the researcher (e.g. the Nanangroe Natural Experiment, Lindenmayer *et al.*, 2008b; Chapter 10);
- **Observational studies**, e.g. the Nest Predation Study (Lindenmayer *et al.*, 1999b);
- **Simulation modelling**, e.g. testing the accuracy of predictions made from models for population viability analysis (Lindenmayer *et al.*, 2003a; Chapter 8);
- The **integration of demographic and genetic methods**, e.g. to better understand dispersal (Peakall and Lindenmayer, 2006; Taylor *et al.*, 2007).

The various research foci and the methods of study have sometimes involved simply describing what patterns of biotic response have occurred in the heavily modified landscape at Tumut. At other times, the focus has been on testing and examining the efficacy of models (Lindenmayer and Lacy, 2002) or testing hypotheses derived from ecological theory, e.g. threshold theory (Lindenmayer *et al.*, 2005b). Clearly, there are many and diverse facets of the research. One of the unifying themes is that all of the work has been focused around the Tumut region, and particularly within and immediately adjacent to the 50,000-ha softwood plantation

---

Box 1.2. The Tumut Study – a true or a natural experiment?

The vast majority of studies of landscape change and habitat fragmentation are observational investigations. These do not use active interventions (e.g. manipulation of sites or other areas) to guide the quantification of biotic responses (or other response variables) to landscape change (reviewed by McGarigal and Cushman, 2002). Typically, observational studies sample a range of existing kinds of sites in a given landscape at one point in time. Such cross-sectional investigations usually lack the replication of site types and/or the 'controls' that characterise true experiments or natural experiments (Lindenmayer and Fischer, 2006).

In contrast to the plethora of observational studies, 'true experiments' that examine the impacts of habitat loss and/or the fragmentation of patches of native vegetation are comparatively rare (Debinski and Holt, 2000). This is perhaps because it is difficult, time-consuming and expensive to manipulate large areas, and difficult to find sufficient replicates, particularly for large patches. Although experiments are a powerful way of investigating the impacts of landscape change on biota, most experiments occur at spatial scales that are too small to provide the practical insights necessary to inform real-world resource management and conservation. While small-scale experiments can reveal interesting findings in some situations (e.g. Golden and Crist, 1999; Lenoir et al., 2003), they can be of limited value for other situations, such as studies of wide-ranging bird and mammal species (Wiens, 1997; Lindenmayer and Fischer, 2006).

'Natural experiments' overlay an experimental design on an ecosystem where change or active manipulation has occurred or is planned (Carpenter et al., 1995). They can be broadly similar to 'true' experiments (Diamond, 1986), but landscape changes are not controlled by the researcher. They usually occur at larger scales than true experiments.

The Tumut Fragmentation Study began in 1995 as a large-scale natural experiment. However, as will become clear throughout this book, the work at Tumut quickly expanded beyond the natural experiment to include many other facets of research, including true experiments, observation studies, integrated demographic and genetic research and simulation modelling.

dominated by stands of Radiata Pine (*Pinus radiata*) that comprise the Buccleuch State Forest.

A guide to using the book

Publication in book form enables more detail to be presented than is typically allowed in standard journal articles, where fierce competition for space invariably means than only the most cursory of background details can be described. References to published papers from the work at Tumut are made throughout the book to direct readers to further details of studies where they are required.

However, some readers might have a particular interest in one particular aspect of the Tumut Fragmentation Study. With this in mind, the book has been divided in a series of logically linked chapters that should enable the reader to navigate to their area of interest quickly.

Chapter 2: The theory contains a short overview of theory relating to landscape change and habitat fragmentation, particularly as it connects to the various major bodies of work at Tumut. It is not an exhaustive treatise on these massive topics, as this has been covered by others (Saunders *et al.*, 1987; Forman, 1995; Lindenmayer and Fischer, 2006; Fischer and Lindenmayer, 2007). However, Chapter 2 provides some background for the chapters that follow.

Chapter 3: The field laboratory describes the characteristics of the study region, and its flora and fauna.

Chapter 4: Setting up the study takes the reader through the many steps associated with the design and implementation of the major cross-sectional study conducted at Tumut, and would be of interest to anyone who is thinking of establishing a large-scale ecological study.

Chapter 5: The core findings discusses the results of the major cross-sectional study for nine groups of animals and plants.

Chapter 6: Patch use describes the results of additional studies undertaken to explore the way in which animals actually used remnant patches, with a focus on the home ranges of arboreal marsupials, bird movements, bird calling behaviour and the breeding behaviour of the Agile Antechinus.

Chapter 7: Theory against data examines how the results from the Tumut Fragmentation Study supported or challenged a number of relevant ecological theories and concepts.

Chapter 8: Testing PVA models with real data continues the discussion on data and theory, with a focus on testing the accuracy of predictions from simulation models for population viability analysis (PVA).

Chapter 9: Genes in the landscape outlines an interesting component of the research at Tumut, which was to explore more deeply the ecological processes of dispersal and movement through integrating genetic and demographic work.

Chapter 10: Refining and extending the research programme describes how deficiencies in the original cross-sectional studies of landscape context effects at Tumut spawned a set of additional investigations that looked at edge effects, nest predation and landscape changes over time.

Chapter 11: Recommendations for plantation managers discusses how the research findings have significant implications for the way in which plantations are managed. Although short, this chapter is valuable because of the rapid expansion of plantation estates around the world and the fact that biodiversity conservation is often considered to be one of the management objectives of these kinds of areas.

Chapter 12: Lessons on running large-scale research studies. This final chapter outlines some ideas for future work. It also highlights some of the many challenges associated with maintaining a large-scale, long-term project. This is an important chapter because many additional lessons can be learned through maintaining a project for a prolonged period of time and some of the most interesting insights emerge only after many years of sustained work.

The chapters in this book vary substantially in length because they cover different amounts of work. Chapter 5, on landscape context effects, for example, is quite long relative to the other chapters because it was a primary focus of the research for many years.

A few caveats

In an age where it is utterly impossible for any one person (or even group of people) to keep up with the exponential increase in the published literature, one aim in putting this book together was to keep it as short as possible. Nevertheless, there was some unavoidable overlap in topics between chapters. This was, in part, because data on particular groups (such as birds) were used in several kinds of studies and to tackle different questions.

This book contains many citations to publications from the work at Tumut (including considerable self-citation). This will frustrate some readers but, given that this is based on published scientific literature, it was unavoidable.

Finally, despite the best efforts of the team of statisticians and ecologists who have been involved in the work at Tumut, the studies completed to date are far from perfect. The perfect ecological study does not exist – even in the minds of the strictest theoreticians! Some limitations of the work are obvious, but others will have undoubtedly been overlooked. Criticisms of the work are welcome so that future research at Tumut (and possibly future editions of this book) might be improved.

2 · The theory: an overview of landscape change and habitat fragmentation

A vast array of topics fall under the umbrella of landscape modification and habitat fragmentation. These include (but are not limited to): conceptual models of landscape cover, habitat loss, habitat degradation, habitat sub-division, edge effects, connectivity, metapopulation dynamics, landscape heterogeneity, threshold effects, extinction cascades, nestedness of biotic assemblages, patch size relationships and landscape genetics. These topics are strongly interrelated and boundaries between them are somewhat artificial (Lindenmayer *et al.*, 2007a) (Figure 2.1).

Many of the individual topics shown in Figure 2.1 have been reviewed in depth over the past two decades, such as connectivity (Bennett, 1990, 1998, Hilty *et al.*, 2006) and edges (Ries *et al.*, 2004; Harper *et al.*, 2005). There also have been major overarching reviews of landscape modification and habitat fragmentation per se (Saunders *et al.*, 1991; Zuidema *et al.*, 1996; Fahrig, 2003; Hobbs and Yates, 2003, Lindenmayer and Fischer, 2006; Fischer and Lindenmayer, 2007). The aim of this chapter is not to produce yet another focused review of a particular topic or of the massively increasing body of work on landscape change and habitat fragmentation; rather, it is to provide brief summaries of topics that were tackled as part of the work at Tumut. These summaries therefore prepare the ground for subsequent chapters that describe the research findings from Tumut.

The 'species-orientated' to 'patterns-based' continuum

Landscape modification and habitat fragmentation can produce a wide range of effects across several spatial scales and levels of biological organ-isation. They can alter many ecological processes, change spatial patterns of vegetation cover in landscapes and influence individual species and assemblages of taxa. Such complexity is reflected by the frequent use of the term 'habitat fragmentation' as an umbrella term for many ecological processes, patterns of vegetation cover and biotic responses that accompany alteration of landscapes by humans (Lindenmayer and Fischer, 2007).

(a)

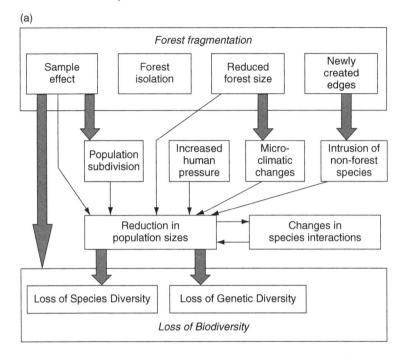

(b)

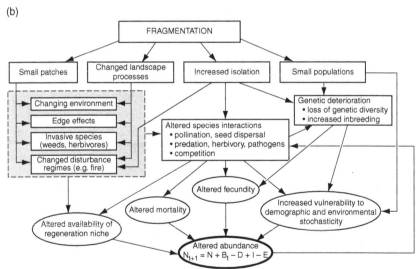

Figure 2.1 Different conceptual models of landscape modification and habitat fragmentation, emphasising links between key topics and areas of research, (a) redrawn from Zuidema *et al.*, 1996; (b) redrawn from Hobbs and Yates, 2003; (c) redrawn from Fischer and Lindenmayer, 2007; (d) redrawn from Lindenmayer *et al.*, 2007a.

(c)

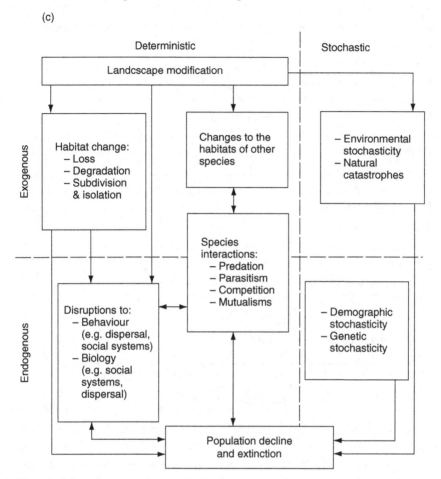

Figure 2.1. (cont.)

Fischer and Lindenmayer (2007) argued that the separate treatment of the different subcomponents of landscape change has led to a range of parallel research paradigms. Two extremes can be identified along a continuum of approaches to understanding the effects of landscape modification on species and assemblages. At one end of the continuum are species-oriented approaches, centred around the response of individual species to landscape change. These approaches are underpinned by recognition of the fact that each species responds individually to landscape change in ways that are broadly related to requirements for nutrients, space, shelter and specific climatic conditions. Interspecific processes such as competition, predation and mutualisms are also important (Fischer and

(d)

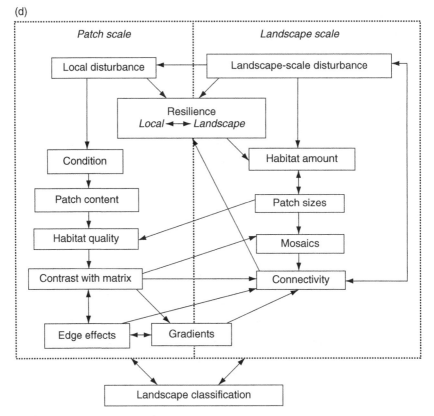

Figure 2.1. (cont.)

Lindenmayer, 2007). At the other end of the continuum are pattern-based approaches in which the focus is on human-perceived landscape patterns and their correlation with measures of aggregate species occurrence (e.g. species richness).

There are strengths and weaknesses of both species-based and pattern-based approaches. The impossibility of studying every individual species in detail limits species-based approaches, whereas the aggregation across species, leading to oversimplification of effects, is a limitation of pattern-based methods (Lindenmayer and Fischer, 2006). This illustrates the complementarity of the two kinds of approaches (Lindenmayer *et al.*, 2007b).

A key part of constructing a framework of approaches for studying the effects on biota of landscape change and habitat fragmentation is the consistent application and interpretation of important terms (Hall *et al.*, 1997) such as those shown in Table 2.1.

Table 2.1. *Key themes and associated terms for use in studies of the impacts of landscape change on biodiversity (redrawn from Fischer and Lindenmayer, 2007; Lindenmayer and Fischer, 2007).*

Theme and term	Definition
Biological organisation and perspective	
Species perspective of a modified landscape	Perception of a landscape by a given (non-human) species; important features include sources of food and shelter, and appropriate climatic conditions
Human perspective of a modified landscape	Perception of a landscape by humans; features include patches of different types of land cover and their spatial arrangement (including native vegetation)
Land cover	
Species perspective	
Habitat	The resources and conditions present in an area that produces occupancy in an area for a particular species[a]
Habitat loss	Loss of habitat for a given species from an area, precluding that taxon from persisting there; i.e. the area becomes non-habitat for that species
Habitat degradation	The reduction in quality or condition of an area of habitat for a given species, thereby impairing the demographics of individuals or populations of that species
Habitat subdivision	Breaking apart of a large area of habitat into several smaller areas
Human perspective	
Native vegetation	The cover of vegetation occurring in an area
Native vegetation loss	Removal of native vegetation (e.g. through land clearing)
Native vegetation deterioration	Reduction in the quality or condition of native vegetation (e.g. relative to a specified benchmark for particular structural features)
Vegetation subdivision	Breaking apart of a single large area of vegetation into several smaller areas
Connectivity	
Species perspective	
Habitat connectivity	Functional linkages between habitat patches for a given species – a species-specific entity
Habitat isolation	Functional separation of habitat patches for a given species – a species-specific entity and the corollary of habitat connectivity

Table 2.1. (*cont.*)

Theme and term	Definition
Human perspective	
Landscape connectivity	Physical linkage (from a human perspective) of areas of native vegetation cover within a landscape
Vegetation isolation	Physical separation of patches of vegetation (as perceived by humans) – the corollary of landscape connectivity
Ecological connectivity	Functional linkages of ecological processes at multiple spatial scales (e.g. trophic relationships, disturbance processes and hydro-ecological flows)

[a]After Hall *et al.* (1997, p. 175)

Based on the concept of species-based and pattern-based approaches along the continuum of approaches to studying landscape change and habitat fragmentation, it is then possible to create a framework that shows how various bodies of work and key topics can be interlinked (Figure 2.2). This framework is, in part, underpinned by the conceptual model of landscape cover (Lindenmayer and Fischer, 2006).

Conceptual models of landscape cover

Effective biodiversity conservation depends, in part, on determining and understanding how landscape change affects organisms (Saunders *et al.*, 1987; Groom *et al.*, 2005). Given such an understanding, conservation biologists are often then called upon to recommend strategies for landscape and conservation planning (e.g. Lambeck, 1999). However, the landscape model they use to characterise landscapes can have a significant influence on the practical conservation recommendations that are made. A brief overview is now presented of some of the landscape models that can be used to conceptualise the effects of landscape change on biodiversity.

A landscape model can be loosely defined as a conceptual tool that provides terminology and a visual representation that can be used to communicate and study how organisms are distributed through space. In theory, landscape models could be applied at many organisational levels – from genes to ecosystems. In practice, the species is the most widely accepted organisational unit in both a scientific and land-management

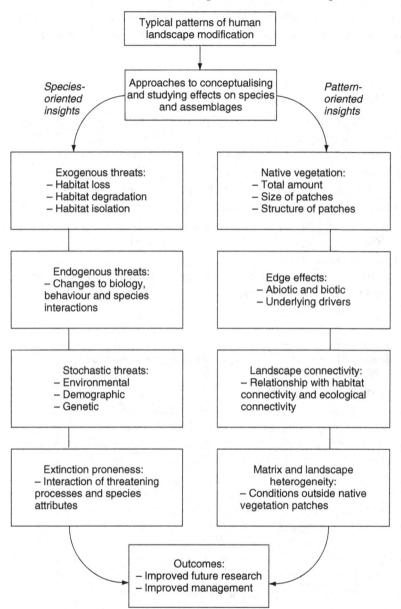

Figure 2.2. Conceptual framework that interlinks species-based and pattern-based approaches to studying the effects of landscape change and habitat fragmentation on biota (redrawn from Fischer and Lindenmayer, 2007).

context (Gaston and Spicer, 2004; Lindenmayer and Burgman, 2005). For this reason, the first example of a landscape model is for individual species.

The landscape contour model

The way that elements of the biota perceive a landscape can be very different from how humans perceive it (Manning *et al.*, 2004). This led Fischer *et al.* (2004) to develop the landscape contour model which incorporates multiple species and their unique habitat requirements, and can characterise gradual habitat change across multiple spatial scales (Figure 2.3). The conceptual foundation comes from Wiens (1995), who suggested viewing landscapes as 'cost–benefit contours', and Lindenmayer *et al.* (1995a), who highlighted how spatially explicit habitat models are similar to contour maps.

Contour maps provide a familiar graphical representation of complex spatial information. The landscape contour model represents the landscape

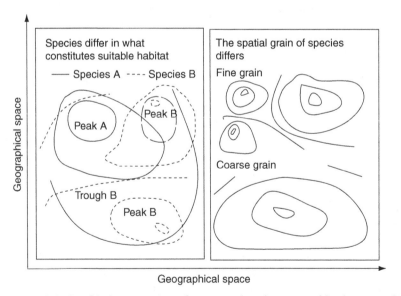

Figure 2.3. Graphical presentation of a contour-based conceptual landscape model. The model recognises gradual changes in habitat suitability through space. Habitat contours are the emergent spatial pattern resulting from a myriad of ecological processes, including availability of suitable food, shelter, climate and sufficient space; as well as competition, predation and other interspecific processes (redrawn from Fischer *et al.*, 2004).

as a map of habitat suitability contours overlaid for different species. The landscape contour model is based on the following premises (after Fischer *et al.*, 2004):

- Habitat is a species–specific concept (Block and Brennan, 1993).
- The spatial grain at which species respond to their environment varies between species (see Kotliar and Wiens, 1990; Manning *et al.*, 2004). This is translated onto a contour map through different contour spacing. Species may have densely spaced contours with many peaks and troughs (fine spatial grain) or widely spaced contours with few peaks and troughs (coarse spatial grain).
- Different spatial resolutions can be used to represent responses at different spatial scales. For example, at the continental scale, the interval between contours can be coarse.
- A habitat contour map for a single species emerges from many ecological processes operating at many spatial scales. The map may not directly reflect the processes causing the pattern of distribution or abundance.
- A contour map does not have a temporal component. Multiple 'snapshots' of contour maps at different times could reflect temporal dynamics such as changing habitat requirements at different stages of a life cycle or changes in habitat suitability with ecological succession (e.g. Palomares *et al.*, 2000).

Limitations of the contour model and single-species approaches

Fundamental to the landscape contour model is the recognition that different species occupy different habitats, and that no two species will respond to landscape change in precisely the same way. While, strictly speaking, this may be true, it poses a difficult challenge to land management (Simberloff, 1998). First, the identity of all species in a given landscape is often not known (Gaston and Spicer, 2004; Lindenmayer and Burgman, 2005). Second, even for those species that are known, it is often not known what constitutes habitat for most of them (Morrison *et al.*, 2006). Hence, in the foreseeable future, detailed studies of the processes affecting individual species' distribution patterns will probably be impossible for the vast majority of species, and will be most likely to occur only for common or charismatic species, or for species of particular conservation concern. This recognition means that some level of generalisation across multiple species is often necessary.

One way to generalise across species is to consider situations where different species have broadly similar habitat requirements, and then focus management efforts on these species. Such groups of species may include 'woodland birds' (Watson *et al.*, 2001, 2005), 'arboreal marsupials' (Lindenmayer, 1997) or 'forest fungi' (Berglund and Jonsson, 2003). The distribution patterns of such groups can be analysed as aggregate measures of species occurrence (such as species richness), e.g. in relation to the size and spatial juxtaposition of forest remnants in a cleared landscape.

Pattern-based landscape models

An alternative to describing landscapes in relation to the habitat requirements of selected species is to describe landscapes from the perspective of humans. Investigating the relationship between human-defined landscape patterns and species occurrence patterns has been a popular, albeit controversial, research area in applied ecology and conservation biology (Haila, 2002). Three landscape models are used particularly frequently, both implicitly and explicitly: the island model, the patch–matrix–corridor model and, to a lesser extent, the variegation model.

The island model

The underlying premise of the island model is that fragments of original vegetation surrounded by cleared or highly modified land are analogous to oceanic islands in an 'inhospitable sea' of unsuitable habitat (reviewed by Haila, 2002). There are three broad assumptions made under the island model. They are:

(1) Islands (or vegetation patches) can be defined in a meaningful way for all species of concern;
(2) Clear patch boundaries can be defined that distinguish patches from the surrounding landscape;
(3) Environmental, habitat and other conditions are relatively homogeneous within an island or patch (Fischer and Lindenmayer, 2007).

Much empirical work has shown that large areas support more species than do smaller ones, and the number of species can often be predicted (albeit rather crudely) from species–area functions (e.g. Arrhenius, 1921; Rosenzweig, 1995; Scheiner, 2003). The theory of island biogeography (MacArthur and Wilson, 1963, 1967) was developed to explain species–area phenomena for island biotas. Part of this theory considers aggregate species richness on islands of varying size and isolation from a mainland source of colonists (Shafer, 1990). The balance between extinction and colonisation, as approximated by island size and isolation, is considered to produce an equilibrium number of species (MacArthur and Wilson, 1967).

The literature on the theory of island biogeography is immense and it is beyond the scope of this chapter to review it (see, e.g., Simberloff, 1988; Shafer, 1990; Whittaker, 1998). Much of it deals with the adaptation of island biogeography theory to reserve design (see Doak and Mills, 1994). However, as a model of landscape dynamics, the island model can fail (e.g. as in Zimmerman and Bierregaard, 1986; Estades and Temple, 1999; Gascon et al., 1999) because:

(1) Areas between patches of remnant vegetation are rarely non-habitat for all species (Daily, 2001; Lindenmayer and Franklin, 2002);
(2) Important interactions between vegetation remnants and the landscapes surrounding them are not accounted for (Janzen, 1983; Whittaker, 1998; Manning et al., 2004).

Despite the inherent problems of the island model, the broad notion of islands in an inhospitable sea has spawned the development of many related theories and concepts (see Rosenzweig, 1995; Whittaker, 1998) that are used widely in work on landscape modification (Doak and Mills, 1994; Pullin, 2002). These include, among many others: wildlife corridors (Bennett, 1998), nested subset theory (Patterson and Atmar, 1986; Chapter 7) and the notion of vegetation coverage thresholds (Andrén, 1994; Chapter 7).

The patch–matrix–corridor model

Rather than conceptualising landscapes as 'islands in a sea of non-habitat', Forman (1995) developed a model in which landscapes are conceived as mosaics of three components: patches, corridors and a matrix (Figure 2.4). The focus is not so much on aggregate species richness, but rather on the geographical composition of landscapes, with the different components having different characteristics, shapes and functions. Forman (1995) defined the three components of his conceptual model as follows:

- **Patches** are relatively homogeneous non-linear areas that differ from their surroundings.
- **Corridors** are strips of a particular patch type that differ from the adjacent land on both sides and connect two or more patches.
- **The matrix** is the dominant and most extensive patch type in a landscape. It is characterised by extensive cover and a major control over dynamics.

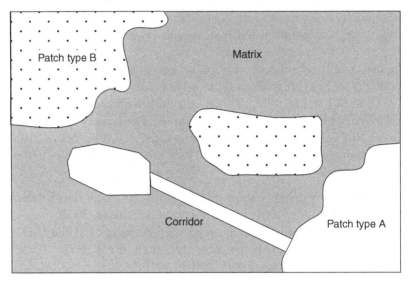

Figure 2.4. A landscape perspective based on the patch–matrix–corridor model (*sensu* Forman, 1995). The fragmented landscape shown here has two patch types (A and B) and an extensive background matrix, and the patches of type A are connected by a corridor.

The name patch–matrix–corridor model is often used for Forman's (1995) conceptualisation of landscape cover patterns. In the patch–matrix–corridor model, the matrix is intersected by corridors or perforated by smaller patches. Patches and corridors are readily distinguished from the background matrix (Forman and Godron, 1986; Kotliar and Wiens, 1990; Figure 2.4). Forman (1995) noted that every point in a landscape was either within a patch, corridor or the background matrix, and that the matrix itself could be extensive to limited, continuous to perforated and variegated to nearly homogeneous.

The patch–matrix–corridor model has been widely adopted in conservation biology. It helps land managers and researchers to translate their ideas into a spatial context. It is an extension of the island model, which often oversimplifies landscapes into areas of habitat and non-habitat. However, the patch–matrix–corridor model also makes some simplifying assumptions. First, it does not generally deal with spatial continua (apart from edge effects; see Laurance *et al.*, 1997; Fischer *et al.*, 2004). Second, it implicitly assumes that a single classification of landscape pattern can work for all species. This can be an important limitation because many organisms do not perceive landscapes in the same way as do humans (Bunnell,

1999a, 1999b; Lindenmayer *et al.*, 2003b). Therefore, patterns of landscape cover seen from a human perspective may not always provide a useful framework for interpreting biotic response to landscape conditions (Manning *et al.*, 2004).

The variegation model

Some authors have objected to the sharp boundaries and discrete classes prevalent in the island model and the patch–matrix–corridor model (Harrison, 1991). In response, McIntyre and Barrett (1992) developed the variegation model (see also McIntyre and Hobbs, 1999). In many landscapes, the boundaries between patch types are diffuse, and differentiating them from the background matrix may not be straightforward. The term 'variegated landscape' was coined to incorporate gradual spatial changes or gradations in vegetation cover (McIntyre and Barrett, 1992; McIntyre, 1994; Manning *et al.*, 2006).

The variegation model was originally proposed for semi-cleared grazing and cropping landscapes in rural eastern Australia. These landscapes are characterised by small patches of woodland and relatively isolated native trees scattered throughout grazing lands (McIntyre and Barrett, 1992; McIntyre *et al.*, 1996). Here, from a human perspective, 'patches' and 'corridors' are difficult to identify among the loosely organised and spatially dispersed trees and ecological communities. For instance, numerous trees scattered across a landscape collectively provide habitat for some species, e.g. for some woodland birds (Barrett *et al.*, 1994; Fischer and Lindenmayer, 2002a, 2002b). The variegation model takes account of small habitat elements that might otherwise be classified as 'unsuitable habitat' in the background matrix (Tickle *et al.*, 1998).

Limitations of pattern-based landscape models

One of the common goals of pattern-based landscape models is to reduce the complexity created by having to analyse every single species in its own right, which would pose insurmountable challenges to landscape management. Some level of simplification of reality in landscape models is necessary, and in fact desirable (Burgman *et al.*, 2005). However, irrespective of the original intentions of the architects of pattern-based landscape models, these models are sometimes used uncritically and can oversimplify ecological patterns. Sometimes, biologists represent landscapes as universally suitable 'habitat patches', contrasting markedly with remaining areas of non-habitat, without carefully assessing whether it is appropriate to aggregate multiple species in this way. Mapping tools (such as GIS: geographical information systems) are sometimes used to define 'habitat patches', assuming that species perceive 'patches' in the same way and at the same scale as do humans (Bunnell, 1999a). Often, approaches based on landscape pattern do not consider the habitat requirements, movement patterns and other important ecological attributes of the organisms of interest.

In some cases, a possible solution to ensure that ecological complexity is not oversimplified is not to use pattern-based landscape models, but instead to apply concepts such as the landscape contour model. In other cases, pattern-based landscape models can be improved by carefully assessing how species should be aggregated. For example, some workers have attempted to overcome the problem of species-specific responses to landscape conditions by classifying species according to their use of a given modified landscape – terms frequently used include 'forest-interior species' (Tang and Gustafson, 1997; Villard, 1998), 'edge species' (Howe, 1984; Bender *et al.*, 1998; Euskirchen *et al.*, 2001) and 'generalist species' (Andrén, 1994; Williams and Hero, 2001). Others have classified patches by the extent to which species use them or by their propensity to provide dispersers (the source–sink concept; Pulliam *et al.*, 1992; Kreuzer and Huntly, 2003). Some concepts distinguish between original vegetation cover and relictual vegetation cover (McIntyre and Hobbs, 1999) or describe the matrix in more detail (Gascon *et al.*, 1999). All these approaches are potentially useful refinements to what might otherwise be an overly simplistic or inappropriate anthropocentric classification of landscape patterns.

The link between single-species and multi-species landscape models

The effects of landscape change can be assessed for a single species, or for multiple species simultaneously. Single-species investigations tend to be more detailed, and tend to demonstrate a reasonable grasp of the ecological processes that limit the distribution and abundance of a given species. In contrast, investigations on multiple species often need to aggregate species into groups and may need to make several assumptions about how landscape patterns are related to a given group of species. A common, but problematic, assumption is that human-defined patches correspond to habitat for a group of species. For example, Watson *et al.* (2001) implicitly assumed that human-defined patches of native vegetation provided essential habitat for woodland birds in the Australian Capital Territory, south-eastern Australia. Later, Watson et al. (2005) demonstrated that the nature of the matrix (i.e. the area between woodland vegetation patches) actually had a major influence on woodland birds, most probably because some species could use the matrix for foraging or breeding, or could readily move through it.

Bearing this in mind, when is it necessary to study every single species, and when is it reasonable to aggregate species and focus on landscape patterns? This dilemma is (often implicitly) a fundamental source of many debates in conservation biology – e.g. about indicator species (Landres et al., 1988; Lindenmayer et al., 2000a), umbrella species (Lambeck, 1997; Roberge and Angelstam, 2004), small reserves (e.g. Gilpin and Diamond, 1980; Simberloff and Abele, 1982; Fischer and Lindenmayer, 2002c) and wildlife corridors (Noss, 1987; Simberloff et al., 1992). The short answer to this dilemma is that there is no simple solution. Rather, it is important to recognise that detailed studies on single species (which often elucidate ecological processes) are complementary to studies that relate human-defined landscape patterns to aggregate measures such as species richness or species composition. Neither approach is inherently superior because both approaches have their own unique limitations (Lindenmayer et al., 2007b).

Single-species approaches are limited mainly in their practical utility, because it will be impossible to study every single species. Multi-species approaches are limited because aggregation usually relies on simplifying assumptions about species co-occurrence patterns. Both single-species and assemblage-based approaches are useful and complementary. However, it is vital to carefully consider when and why it is appropriate to select a particular approach (Lindenmayer et al., 2007b).

Research at Tumut

The work at Tumut spawned a range of studies of different conceptual models, especially as they relate to the outcomes of investigations of landscape context effects for different groups of animals and plants (see Chapters 5 and 7). In particular, the landscape contour model (Fischer et al., 2004; see above) was developed, in part, as a consequence of the findings derived from work conducted at Tumut.

Landscape context and landscape heterogeneity

In some situations, the spatial distribution of assemblages can be described well by the patch–matrix–corridor model where the matrix is considered to be the largest and often most highly modified patch type in a given landscape (*sensu* Forman, 1995). In these circumstances, the matrix can have many important roles. For example, it can affect the population dynamics of individual species, including population recovery after disturbance (Lamberson *et al.*, 1994; Turner *et al.*, 2003), the exchange of individuals between patches (Leung *et al.*, 1993; Schmuki *et al.*, 2006a) and occupancy rates of vegetation patches (Villard and Taylor, 1994).

In many modified landscapes, the notion of a matrix that is inhospitable to most species is not useful (Barrett *et al.*, 1994; Lindenmayer *et al.*, 2003b). Examples where the matrix provides habitat for a range of species come from both forestry and agricultural landscapes. Despite the potential value of the matrix as habitat, it is important to recognise that not all species will survive in all types of matrix (Renjifo, 2001). A given species' survival in the matrix depends on the interaction of its particular requirements with the environmental conditions in the matrix (Laurance, 1991; Andrén, 1997). In addition to providing permanent habitat, the matrix can be a movement conduit for some organisms, thus reducing the negative effects of habitat isolation (Joyal *et al.*, 2001; Bender and Fahrig, 2005). A final and related key role of the matrix is to provide the ecological context for native vegetation patches located within it. Conditions in the matrix can have an important effect on which species are present in patches of native vegetation. Landscape context effects become particularly noticeable when species composition is compared between patches composed of similar vegetation types, but located in different matrix types

Some landscapes are so diversely patterned that the delineation of patches within a dominant background matrix (which itself may be highly non-uniform) becomes difficult or arbitrary. In particular, patch boundaries defined from a human perspective may be irrelevant for many organisms, including some of conservation interest. An alternative way to conceptualise highly heterogeneous landscapes is to apply the landscape contour model (Figure 2.3), which recognises that each species has its own individualistic patterns of distribution. In such heterogeneous landscapes, different vegetation types occur side by side and structural complexity may also be high, and so many different types of niches are available that can be used by different organisms. Heterogeneous landscapes therefore often support more species than otherwise similar, but less heterogeneous, landscapes.

Research at Tumut

The major focus of work at Tumut involved the quantification of land-scape context effects for an array of biotic groups (Chapter 5). Relationships between biotic responses and landscape heterogeneity were identified as part of research on landscape indices and the results of that work are summarised in Chapter 7.

Threshold responses to native vegetation cover

Different effects of landscape modification can interact (Figure 2.1, Figure 2.2). When these effects reinforce one another, they can trigger cascades of ecosystem change that can be difficult or impossible to reverse. Such cascades are sometimes also referred to as 'regime shifts'. Regime shifts occur when interrelationships between key variables in an ecosystem change fundamentally – they can be thought of as transitions, where an ecosystem 'flips' from one state to another (Scheffer et al., 2001; Folke et al., 2004; Walker and Meyers, 2004).

Thresholds may exist for the amounts of particular kinds of vegetation cover in a landscape. With and King (1999) broadly defined thresholds as abrupt, non-linear changes that occur in some measure (such as the rate of loss of species) across a small amount of habitat loss. When these thresholds are breached, sudden changes in species abundance or ecosystem processes may occur, leading to changes in system state or regime shifts (Folke et al., 2004). For example, below a critical amount of native vegetation cover, the loss of species and populations is greater than can be predicted from a smooth relationship with native vegetation cover alone (Rolstad and Wegge, 1987; Andrén, 1994, 1999; Enoksson et al., 1995; With and Crist, 1995; With and King, 1999).

Hypothetically, thresholds are more likely to be crossed and regime shifts are more likely to occur when levels of particular kinds of vegetation cover are low (e.g. 10–30%; Radford et al., 2005). A critical aspect of landscape conservation must be to avoid reaching such levels. The threshold concept has also been used to recommend benchmarks for land management (McAlpine et al., 2002a), such as for native vegetation retention on farms (Barrett, 2000; McIntyre et al., 2000). Thus, the possibility of a broad rule of thumb about how much native vegetation is needed to avoid regime shifts – e.g. a 30% threshold – has received considerable attention both by scientists and workers in the policy arena (e.g. see review by Huggett, 2005).

The '30 percent rule' for threshold effects in habitat loss has been examined for individual species, assemblages (Parker and Mac Nally, 2002; Drinnan, 2005; Radford *et al.*, 2005; Sasaki *et al.*, 2008), and also broader measures of ecosystem integrity (e.g. tree health; see McIntyre *et al.*, 2002). However,

(1) It is difficult to identify critical change points or thresholds and to anticipate (and prevent) regime shifts before these occur (Groffman *et al.*, 2006);
(2) Thresholds may not exist for some measures such as aggregate species richness because of contrasting responses of individual species (Lindenmayer *et al.*, 2005b). Nevertheless, the key point is to recognise the possibility of non-linear responses to landscape modification and the existence of critical zones in which rapid change occurs.

Research at Tumut

In respect of the research at Tumut, quantitative analyses of threshold effects in the total amount of native vegetation cover were completed using data gathered for birds and reptiles. The results of that work are summarised in Chapter 7.

Patch size relationships

The particular use of the term 'habitat', coupled with how a landscape is classified and mapped, will determine what constitutes a 'patch' – a patch of habitat for a given species, or a patch of vegetation of a particular type. In both cases, larger patches have been considered critical (reviewed by Hanski, 1994a). This is because of relationships between patch size and

(1) The size and extinction proneness of populations of individual species;
(2) Species richness;
(3) Many other factors (e.g. immigration rates, disturbance sizes, vegetation diversity).

While large patches are important, many studies have shown that the ecological value of small and medium-sized patches can be considerable (Turner, 1996). In addition, patch size is relative: what constitutes a large patch of habitat for a species of beetle may be a small patch for a species of bird or mammal (Wiens *et al.*, 1997).

Many studies have focused on individual patches or sites within patches, but patch size effects cannot be divorced from other pivotal issues such as the occurrence or role of ensembles of patches or patch mosaics – a topic that still remains poorly understood (Bennett *et al.*, 2006). Mosaics of different patches (in different condition and characterised by different internal structure) are likely to be important (Law and Dickman, 1998), as shown from research in fire dynamics (e.g. Parr and Andersen, 2006). However, it is not straightforward to determine which mosaic is the most appropriate one to maintain (Bradstock *et al.*, 2005), particularly given the dynamic temporal aspects of mosaics.

Research at Tumut

Patch size relationships for various groups have been studied in several research projects at Tumut, including those on:

- Landscape context effects (Chapter 5);
- Landscape indices (Chapter 7);
- The peninsula effect (Chapter 7);
- Nestedness (Chapter 7);
- Modelling population persistence (Chapter 8);
- Population recovery following the experimental removal of small mammals (Chapter 9).

In addition to research on patch size relationships, complementary work has examined how several individual species use patches of different sizes (e.g. for movement; Chapter 6).

Habitat subdivision, habitat isolation and metapopulation dynamics

The modification of landscapes by humans often results in the subdivision of remaining areas of a given species' habitat. Habitat subdivision involves breaking a habitat patch into two or more smaller patches (Fahrig, 2003). This is congruent with various dictionary definitions of fragmentation; that is, 'a breaking apart'. The smaller size of the remaining habitat patches resulting from habitat subdivision may result in these patches being unable to support viable populations (Armbruster and Lande, 1993). In addition to breaking larger single patches into smaller ones, habitat subdivision can result in the remaining habitat patches becoming more isolated from one another.

Some species have a naturally patchy distribution. Particularly good examples are plants and animals that are largely confined to rocky outcrops (Hopper, 2000; Schilthuizen *et al.*, 2005) or those that live on mountain tops (Mansergh and Scotts, 1989) and in freshwater ponds (Sjorgen-Gulve, 1994). However, many other species whose natural distribution is more continuous are negatively affected by habitat subdivision.

For plants, habitat isolation may limit the movements of propagules such as spores, pollen and seeds (Duncan and Chapman, 1999; Cascante *et al.*, 2002). For animals, habitat isolation may impair movement at several spatial and temporal scales. This interruption to dispersal can reduce the genetic size of populations through impaired patterns of gene flow (Leung *et al.*, 1993; Epps *et al.*, 2005). By affecting patterns of dispersal between patches, habitat isolation can have significant effects on the occupancy of otherwise suitable habitat patches (Villard and Taylor, 1994; Pither and Taylor, 1998) and on population persistence (Mills and Allendorf, 1996), thereby contributing to the extinction proneness of species (Angermeier, 1995).

Habitat isolation may shift a formerly contiguous and interacting population into a series of loosely connected subpopulations, i.e. a metapopulation. A metapopulation is defined as 'a set of local populations which interact via individuals moving between local populations' (Hanski and Gilpin, 1991, p. 7).

The term metapopulation is sometimes applied to all species distributed across a number of vegetation remnants (Arnold *et al.*, 1993). However, true metapopulations are defined by dynamic properties, including the following (after Hanski, 1997, 1999):

- Suitable habitat must be restricted to particular patches that can be differentiated from the surrounding unsuitable matrix.
- The populations in most patches must be at risk of extinction at some stage.
- There must be inter-patch dispersal and colonisation, but not so much that the local dynamics of populations within a patch are not synchronous with the dynamics of populations in other patches.

Patchily distributed populations of a species do not always conform to a true metapopulation structure (Bradford *et al.*, 2003). For example, the areas between vegetation patches may be so hostile that they preclude movement, so that populations are confined to isolated habitat patches (Sarre *et al.*, 1995). At the other extreme, species in landscapes that appear patchy from a human perspective may also not be distributed as metapopulations if areas between patches are sufficiently suitable to facilitate frequent movements between them (Price *et al.*, 1999), or if conditions allow organisms to forage or live in the areas between patches (Tubelis *et al.*, 2004).

If a species has a metapopulation structure, there should be a characteristic spatial distribution pattern in which locations close to occupied patches are more likely to be occupied than more distant ones (Hanski, 1994a; Smith, 1994; Koenig, 1998). The metapopulation concept is most useful when successful inter-patch dispersal is infrequent and migration distances are limited (Hanski and Simberloff, 1997).

Research at Tumut

Habitat isolation and metapopulation dynamics were at the core of four research projects at Tumut. The first was a study of small mammal recovery following experimental population reduction (Chapter 9). The second was a series of studies of genetic variability in native mammals (Bush Rat, Agile Antechinus and Greater Glider) (Chapter 9). The third was population modelling and comparisons between predicted and actual patch occupancy for a range of mammal and bird species (Chapter 8). Finally, analyses of patterns of spatial autocorrelation of patch occupancy were completed as part of landscape context work on birds (Chapter 5). This was designed to test whether field values for patch occupancy were congruent with the patterns expected if metapopulations were occurring.

Connectivity

Connectivity relates to the ability of species and ecological resources and processes to move through landscapes, not only in the terrestrial domain, but also in aquatic systems and between the two. Connectivity, and in particular the value of corridors, has been much debated. Some debates about connectivity stem from the term being too broadly conceived, rendering it difficult to use in practice, and different interpretations of terms. Lindenmayer and Fischer (2007) have suggested that some of the controversy might be avoided by making a careful distinction between:

(1) Habitat connectivity or the connectedness of habitat patches for a given taxon;
(2) Landscape connectivity or the physical connectedness of patches of native vegetation cover as perceived by humans;

(3) Ecological connectivity or connectedness of ecological processes at multiple spatial scales.

Lindenmayer and Fischer (2007) further noted that although the three connectivity concepts are interrelated, they are not synonymous. In some circumstances, habitat connectivity and landscape connectivity will be similar (Levey *et al.*, 2005). In others, habitat connectivity for a given species will be different from the human perspective of landscape connectivity.

Complexity, issues and interrelationships

Connectivity remains one of the most difficult areas of landscape conservation. Measuring connectivity is not straightforward and the metrics used can be highly problematic (Tischendorf and Fahrig, 2000a, 2000b). Habitat connectivity and other forms of connectivity are hard to study because they are interrelated with the notoriously difficult area of dispersal biology (Keogh *et al.*, 2007). Nevertheless, a better understanding of connectivity is urgently required, given the impending effects of rapid climate change on species distributions and the potential for shifts in species ranges to be blocked by human modification of landscapes.

Although most ecologists agree about the importance of various kinds of connectivity, disagreement arises when connectivity is equated simply with corridors or linear strips of native vegetation that link vegetation patches. The supply of corridors is just one of several approaches to providing connectivity for some species and ecological processes (Levey *et al.*, 2005). The simplicity of the corridor concept and the relative ease with which corridors can be implemented in planning exercises can lead to a failure to consider the connectivity function of the surrounding areas (Hannon and Schmiegelow, 2002). This emphasises that the topic of connectivity cannot be readily divorced from others such as the amount of native vegetation cover in the landscape and the value of that cover as habitat for particular species (Fahrig, 2003).

Research at Tumut

The most detailed work on connectivity at Tumut involved coupling demographic and genetic work on mammal and beetle populations within patches of remnant native vegetation and contiguous areas of native forest. This research is summarised in Chapter 9.

Nestedness

An assemblage distributed across a number of discrete sites is considered nested when the taxa present at relatively species-poor sites are subsets of those present at progressively more species-rich sites (Patterson and Atmar, 1986). Whittaker (1998) likens the concept of nestedness to that of 'Russian dolls', in which each doll has a smaller one inside it. Nested subset theory stems from observations of the composition of assemblages on islands (Darlington, 1957) and mountain tops (May, 1978). More recently, nested assemblages have also been observed in landscapes altered by human activities (e.g. Cornelius et al., 2000), and in relatively unmodified patchy landscapes or regions (e.g. Patterson and Brown, 1991).

A review of over 150 datasets from a range of climatic regions (including studies on vertebrates, molluscs, plants and arthropods) concluded that nestedness was a common phenomenon for a wide range of assemblages (Wright et al., 1998). Worthen (1996) considered that the presence or absence of nestedness was a useful and fundamental descriptor for community composition. Although many assemblages are significantly nested (Atmar and Patterson, 1995; Wright et al., 1998), it is important to note that very few assemblages are perfectly nested (sensu Figure 2.5). That is, although many assemblages can be arranged as roughly triangular species × sites matrices, most exhibit some unexpected presences ('outliers') or absences ('holes'; see Cutler, 1991). The difference between perfect and statistically significant nestedness can have important implications for the use of nested subset theory in a conservation context, including in human-modified landscapes (Fischer and Lindenmayer, 2005a; see Box 7.1 in Chapter 7).

In significantly nested assemblages, the occurrence sequences of species across sites are not random (Schoener and Schoener, 1983; Simberloff and Levin, 1985). Rather, some species occur in nearly all sites (e.g. species 'A' and 'B' in Figure 2.5), whereas others only occur in species-rich sites (e.g. species 'D' and 'E' in Figure 2.5). This non-random pattern suggests that one or several mechanisms structure the assemblage in a predictable way. Three ecological mechanisms are considered to be particularly important. These are selective extinction, selective immigration and habitat nestedness.

Not all assemblages are nested. For example, Graves and Gotelli (1993) studied mixed-species flocks of birds in Amazonia, and found that ecologically similar congeneric species often replaced one another in different

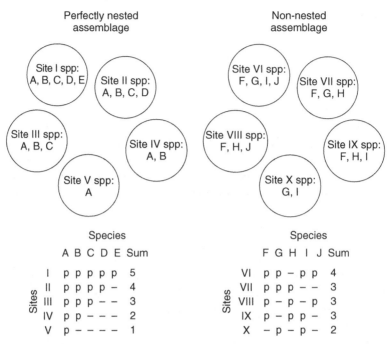

Figure 2.5. Examples of two assemblages containing species A, B, C, D, E and species F, G, H, I, J. For a given species, 'p' denotes its presence and '−' indicates its absence at a given site. The left assemblage is perfectly nested, i.e. the species present at species-poor sites are subsets of those present at progressively richer sites. The right assemblage is not nested (redrawn from Lindenmayer and Fischer, 2006).

flocks. Thus, species were not added or lost from the assemblage in a predictable sequence, but rather different sets of species occurred in different flocks. Because of the resulting pattern in species × sites matrices, this type of assemblage is often referred to as a 'chequerboard'.

Patterson and Brown (1991) identified three necessary (but not automatically sufficient) preconditions for nestedness to occur. First, sites need to have a common biogeographical history. Second, contemporary environments at all sites need to be broadly similar. Third, niche relationships of species need to be organised hierarchically. The first two conditions are important to ensure that all sites have a common pool of potentially co-occurring species. The third condition is important to ensure that the sequence of presences of species at a series of sites varies in a predictable and monotonic way (Bruce Patterson, personal communication). Departures from these conditions can mean that nestedness is weakened or absent.

Research at Tumut

A series of studies examined the data gathered for reptiles, birds and mammals at Tumut for evidence of nestedness. The results of that work are summarised in Chapter 7.

Edges

A common change in landscape pattern resulting from landscape alteration is the increase in the length of boundaries or edges between remaining patches of original vegetation and the surrounding matrix. Edge effects can occur at such boundaries and can have profound impacts on plant and animal assemblages (Laurance, 2000; Siitonen et al., 2005). Empirical studies have shown that sharp boundaries between patches influence diverse biotic and abiotic processes (Ries et al., 2004; Harper et al., 2005).

One way of classifying edges is by their origin – natural or human-derived (Luck et al., 1999). Edges usually occur naturally at the interface of two ecological communities. Such natural boundaries are sometimes termed ecotones and can support distinctive plant and animal communities (Witham et al., 1991).

In modified landscapes, many edges are created by humans. Examples are the boundaries between recently clear-cut forest and adjacent unlogged stands (Chen et al., 1990) or between cropped areas and remnant native vegetation (Sargent et al., 1998). Edges can be 'soft', where the transition between different types of patches is gradual; or 'hard' at boundaries with marked contrasts in vegetation structure and other features, e.g. clear-felled/unlogged forest (Forman, 1995).

Edge effects refer to changes in biological and physical conditions that occur at an ecosystem boundary and within adjacent ecosystems (Wilcove, 1985). The ecological edge effects that result from a disturbance are the outcome of interactions between the kind and intensity of the disturbance event and the ecological dynamics within the adjacent, undisturbed environment.

In a major review of edge effects in forest landscapes, Harper et al. (2005) classified edge effects as either primary responses that arise directly from edge creation or secondary responses that arise indirectly as a result of edge creation. Primary responses include structural damage to the vegetation, disruption of the forest floor and soil layer, altered nutrient cycling and decomposition, changed evaporation and altered pollen and seed dispersal (Harper et al., 2005). Secondary or indirect responses (that flow from primary responses) include patterns of plant growth, regeneration, reproduction and mortality, and are manifest as altered patterns of vegetation structure and species composition (Harper *et al.*, 2005). The intensity of edge effects or the area of a patch subject to significant edge influence depends on the parameter of interest (Laurance *et al.*, 1997).

Another way of classifying edge effects is by the impacts they have on climate and abiotic processes (e.g. altered wind penetration) or on biota (e.g. altered levels of predation). These two broad kinds of effects are termed abiotic edge effects and biotic edge effects.

Abiotic edge effects

Edges may experience microclimatic changes such as increased temperature and light, or decreased humidity, that extend tens or hundreds of metres from an edge, depending on the environmental variable, the physical nature of the edge and weather conditions (reviewed by Saunders *et al.*, 1991; Laurance *et al.*, 1997).

Biotic edge effects

Edge environments can affect reproduction, growth, seed dispersal and mortality in plants (Hobbs and Yates, 2003). For example, Chen (1991) observed increased reproduction and growth of surviving mature trees in old-growth forests bordering recently clear-felled areas in the Pacific Northwest of the USA. Similarly, weed invasion is a major biotic edge effect in many heavily disturbed landscapes (Brothers and Spingarn, 1992; Krebs, 2008). Altered microclimatic conditions at edges may make conditions particularly favourable for the growth of non-native plants (Honnay *et al.*, 2002). Regenerating vegetation and patch edges often experience a seed rain of environmental weeds and other introduced plants that are frequently better adapted to exposed and disturbed environments (Janzen, 1983).

A wide range of studies have demonstrated the existence of altered animal community composition at edges (e.g. Hansson, 1998), including assemblages of invertebrates (e.g. Magara, 2002). Some vertebrates avoid forest edges and are classified as 'interior forest' species (Gates and Gysel, 1978). Some forest birds also fall into this category (Terborgh, 1989), in part because arthropod densities may be lower near edges (Zanette and Jenkins, 2000) or because foraging efficiency may be impaired by edge conditions (Huhta *et al.*, 1999). Another common reason for species avoiding edges is that higher abundances of predators and parasites at edges can have negative impacts on prey and host species near edges (Chalfoun *et al.*, 2002). For example, bird populations inhabiting edge environments can experience lower rates of successful pairings and impaired breeding success compared with populations of the same taxa occupying other parts of a landscape (Chalfoun *et al.*, 2002). Frequently studied predators in this context include corvids, small mustelids and other small mammals (e.g. the Red Squirrel, *Tamiasciurus hudsonicus*), as well as larger carnivores such as Foxes (*Vulpes* spp.) and Badgers (*Meles meles*). The most frequently studied nest parasite is the Brown-headed Cowbird (*Molothrus ater*) in North America (Zanette *et al.*, 2005).

Discussions of edge effects are intimately linked with other key themes in landscape ecology and conservation biology. An example is the relationship between edge effects and the amount and spatial configuration of particular vegetation types in a landscape (Bayne and Hobson, 1997). Similarly, a particular classification of a landscape and the map generated from it will influence where and how edges are perceived to occur. Edges are usually considered from a human perspective and scale (Bunnell, 1999b), but edges and edge effects can exist at many scales (Laurance *et al.*, 2001), and the way in which organisms perceive and respond to edges will often differ from that of humans. Hence, simple categorisations such as 'edge' or 'interior' species will often provide inadequate descriptors of complex responses. The assumption that edges and edge effects are important is also linked to the perpetuation of a patch-based conceptualisation of landscapes. Patch-based classifications of complex landscapes may be artefacts of the mapping process. A large literature demonstrates that both biotic and abiotic variables change as continuous functions from the interior of one patch to the interior of the adjoining patch (Sisk *et al.*, 2002). Gradient approaches that complement patch-based conceptualisations of landscapes may help to foster the continued development of ideas about how landscape patterns, including edges, can influence ecological processes (Fischer *et al.*, 2004).

There is much ecological variation in response to different kinds of edges – among taxa, between vegetation types and between regions. Although the magnitude of responses to particular edge effects may differ, often the nature of the effect (i.e. positive or negative) will not. Mechanistic approaches based on the strength of habitat associations and resource availability may help to clarify the nature and strength of edge effects and provide a foundation for improved predictive models – at least for biotic edge effects. Ries *et al.* (2004) and Harper *et al.* (2005) provide useful conceptual approaches in this regard.

Research at Tumut

Three studies directly explored edge effects at Tumut. First, edge effects were explored specifically as part of landscape context effects and invasive plant occurrence in eucalypt remnants (Chapter 5). Second, a nest predation experiment explored relationships between the loss of quail eggs from artificial nests across different kinds of boundaries (Chapter 10). Third, the length of boundaries was a key covariate in work on relationships between the occurrence of particular species and landscape indices (Chapter 7). Boundary effects also formed part of the work on bird occurrence in the woodland remnants in the Nanangroe Natural Experiment (Chapter 10).

Summary

Landscape change and habitat fragmentation are massive and multifaceted topics accompanied by a vast and rapidly increasing literature. While this chapter has summarised many topics that collectively fall under the umbrella of landscape change and habitat fragmentation, other topics such as habitat loss and habitat degradation have not been reviewed because they were not studied in detail at Tumut. This was because much of the work has been cross-sectional and observational, with 1985 being the last

Table 2.2. *Some of the different kinds of landscape change and habitat fragmentation research conducted at Tumut, and the chapters in which they are described in this book.*

Topic	Chapter/s
Conceptual landscape models	7
Landscape context	5
Landscape heterogeneity	7
Threshold effects	7
Patch size relationships	5, 6, 7, 8, 9
Habitat isolation and metapopulation dynamics	5, 8, 9
Connectivity	9
Nestedness	7
Edge effects	5, 7, 10

year of forest clearing (= habitat loss for some species) to establish stands of plantation pine.

Many different kinds of studies of several facets of landscape change and habitat fragmentation were completed at Tumut. These are highlighted in Table 2.2.

3 · *The field laboratory: the Tumut study area and the vertebrate animals it supports*

This chapter outlines some of the key features of the study region, including types and patterns of vegetation cover and attributes of some of the flora and fauna. More detail is provided on those species targeted for detailed research.

Geology and climate

The work at Tumut was focused in and around the Buccleuch State Forest, which lies 100 km west of Canberra in southern New South Wales (NSW), south-eastern Australia (Figure 3.1). The total area under investigation encompasses ~100,000 ha and its midpoint is 148°40′E, 35°10′S.

The Tumut study area is characterised by an elevation gradient of 400–1,200 m above sea level. The terrain varies from mountainous and steep to undulating. Permanent streams flow through the valleys of the region, and there are also semi-permanent wetlands and soaks (Parris and Lindenmayer, 2004).

Two climate domains, as defined by the Köppen Climate Classification System, can be found at Tumut (Dick, 1975), with rainfall distributed evenly across the year and either cool or hot summers. A detailed analysis using the climate projection system BIOCLIM (Nix and Switzer, 1991) shows that the estimated mean annual temperature varies from 9.6 to 13.8°C and the annual mean precipitation from 785 to 1,385 mm.

The geology underlying Buccleuch State Forest and its immediate surrounds is dominated by granites interspersed with limited Tertiary alluvial and volcanic deposits. In addition, there is a small area of Cainozoic sediments confined to the central part of the plantation and a belt of intrusive igneous rocks spanning the plantation/native forest boundary in the southern part of the study area. A narrow seam of serpentine rocks runs approximately north–south along part of the study area's western edge.

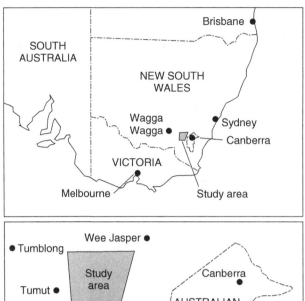

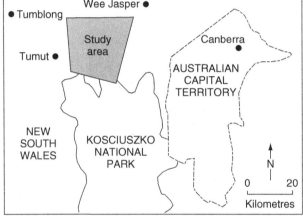

Figure 3.1. The general location of the study area at Tumut.

Plantation vegetation cover

The Buccleuch State Forest is a 50,000 ha plantation of exotic softwoods established on areas that formerly supported native eucalypt forest. The vast majority of softwood stands are Radiata Pine (*Pinus radiata*), although limited areas of other softwood trees species such as Douglas Fir (*Pseudotsuga menziesii*) have been established.

Extensive clearing of native forest for Radiata Pine commenced in the mid 1930s and continued for 50 years, until changes to forest clearing legislation resulted in new areas of softwood plantation being established on grazing land that had been previously cleared or at least partially cleared (Figure 3.2).

(a)

(b)

Figure 3.2. Clearing of native vegetation cover to establish stands of plantation pine at Tumut (photos by David Lindenmayer).

The rotation time for plantation management at Tumut is approximately 25–30 years, depending on site productivity. Areas targeted for plantation establishment are deep-ripped with a bulldozer and young Radiata Pine seedlings are planted into mounds of earth (Figure 3.3).

Stands are thinned within approximately 12 years of establishment and then again at approximately 20 years. Depending on site quality, stands may be thinned at about 25 years and then clear-felled at 30 years. Alternatively, stands may be clear-felled at 25 years.

(a)

(b)

Figure 3.3. (a) Mounded soil into which young Radiata Pine seedlings are planted; (b) Machine-logging of plantation Radiata Pine; (c) Steep slope harvested by cable-logging (photos by David Lindenmayer).

(c)

Figure 3.3. (cont.)

The oldest parts of the Buccleuch State Forest are now in their third rotation, whereas only one rotation has taken place in areas which were the last to be converted from native forest to plantation in the mid 1980s. Almost all logging is by machine, except on steep terrain, where cables are employed – with trees being initially cut by hand before being attached to aerial wires.

After clear-felling at the end of a logging rotation, logging slash is heaped and burned prior to bulldozer ripping and mounding to prepare the ground for planting to commence a new rotation of Radiata Pine stands. A small number of native trees or understorey plants may survive these procedures (Smith, 2006), although the ground cover and understorey layers of plantation stands in almost all parts of Buccleuch State Forest are largely dominated by the introduced but invasive Blackberry (*Rubus fruticosus*) (Lindenmayer and McCarthy, 2001).

Native vegetation cover

Many areas of native vegetation escaped clearing as part of plantation establishment in the Buccleuch State Forest between 1932 and 1985. A total of 192 patches of remnant native eucalypt forest occur within the boundaries of the plantation. These areas of native vegetation vary in size

Figure 3.4. Aerial view of eucalypt patches embedded within the Radiata Pine matrix at Tumut (photos by David Lindenmayer).

(0.1–124 ha), shape, time since isolation and a wide range of other characteristics (see Figure 3.4).

They are dominated by stands of eight different eucalypt species. These are Narrow-leaved Peppermint (*Eucalyptus radiata*), Mountain Swamp Gum (*Eucalyptus camphora*), Red Stringybark (*Eucalyptus macrorhynca*), Apple Box (*Eucalyptus bridgesdiana*), Ribbon Gum (*Eucalyptus viminalis*), Mountain Gum (*Eucalyptus dalrympleana*), Snow Gum (*Eucalyptus pauciflora*) and Broad-leaved Peppermint (*Eucalyptus dives*).

Other minor tree species that are relatively uncommon in the study area include: Eurabbie Blue Gum (*Eucalyptus bicostata*), Alpine Ash (*Eucalyptus delegatensis*), Brown Barrel (*Eucalyptus fastigata*), Blakely's Red Gum (*Eucalyptus blakelyi*), Candlebark (*Eucalyptus rubida*), Brittle Gum (*Eucalyptus mannifera*) and Long-leaved Box (*Eucalyptus gonioclayx*).

The forests in the Tumut region can support relatively tall understorey plants (up to 10 m in height), with dominant elements including such species as Silver Wattle (*Acacia dealbata*), Broad-leaved Hickory Wattle (*Acacia falciformis*), Blackwood (*Acacia melanoxylon*) and Cherry Ballart (*Exocarpus cupressiformis*), as well as 'wildling' Radiata Pine trees that have invaded the forest from the neighbouring plantation.

In addition, to the major species of overstorey and understorey trees, the flora of the study area includes at least an additional 188 species of native vascular plants (Smith, 2006). These include rare and threatened species such as the Tumut Grevillea (*Grevillea wilkinsonii*). Non-vascular plants are also species-rich; Pharo *et al.* (2004) identified 45 species of mosses and 13 liverworts.

The range of types of native vegetation that occur in the patches of remnant native vegetation in the plantation are also found in the extensive areas of native forest that characterise the areas to the north, east and south of Buccleuch State Forest. These encompass the Bondo and Bungongo State Forests, where limited wood production by native forest harvesting occurs, and the Brindabella and Kosciuszko National Parks, which are managed primarily for nature conservation and human recreation. As we discussed in detail in Chapter 4, the similarity of the native vegetation within and adjacent to the plantation is a key feature of the experimental design for the Tumut Fragmentation Study.

Disturbance

Logging, either repeated thinning or clear-felling, is obviously the main form of human disturbance in the Buccleuch State Forest. The primary form of natural disturbance in the study area is wildfire (Cary, 2002), although localised damage to both native forest and plantation stands can occur as a result of windstorms and snowfalls. In December 2006 and January 2007, a wildfire burned extensive areas of the Buccleuch State Forest and killed over 10,000 ha of Radiata Pine. It also burned patches of remnant native forest within the plantation as well as more extensive areas of continuous native forest at the plantation boundary. The wildfire triggered post-fire salvage logging to recover economic losses from killed Radiata Pine trees and to prepare the ground for replanting with pine seedlings (Figure 3.5).

Fauna of the Tumut region

Vertebrates have been the focus of much of the research at Tumut region. Several earlier studies helped document some of the fauna of the area. This previous work was dominated by mammal research and surveys (Smith, 1969; Tyndale-Biscoe and Smith, 1969a, 1969b; Marples, 1973; Gall, 1982; Rishworth *et al.*, 1995a, 1995b). One limited survey documented the occurrence of reptiles in an area near to Tumut (Caughley and Gall, 1985), but no formal studies surveyed for birds, although location records were gathered as part of the Australian bird atlas (Blakers *et al.*, 1984; Barrett *et al.*, 2003).

(a)

(b)

Figure 3.5. Post-fire salvage logging at Tumut following wildfires in 2006/2007 (photos by David Lindenmayer).

In the remainder of this chapter, some features of several of the key vertebrate groups which have been the focus of much of the work at Tumut are outlined. This information is provided to help guide readers in their understanding of the animal responses to landscape change that are described in subsequent chapters. Of course, this section of the book is highly summarised because most vertebrate assemblages are diverse and species-rich. For example, they include:

- More than 90 species of birds from 34 families (see Appendix 2; Lindenmayer *et al.*, 2007c);
- Eight species of arboreal marsupials and approximately 20 species of terrestrial mammals, about half of which are introduced (Lindenmayer *et al.*, 1999c);
- Over 20 species of reptiles (Fischer *et al.*, 2005);
- Eight species of frogs (Parris and Lindenmayer, 2004).

Birds

Birds are by far the most species-rich vertebrate assemblage in the Tumut region (see Appendix 2; Lindenmayer *et al.*, 2007c). It is a largely forest-dominated assemblage and includes such well-known and highly charismatic taxa as the Superb Lyrebird (*Menura novaehollandiae*) and several psittacid and cacatuid parrots − Crimson Rosella (*Platycercus elegans*), Eastern Rosella (*Platycercus eximus*), King Parrot (*Alisterus scapularis*), Galah (*Cacatua roseicapilla*), Gang-gang Cockatoo (*Callopcephalon fimbriatum*), Sulphur-crested Cockatoo (*Cacatua galerita*) and Yellow-tailed Black Cockatoo (*Calyptorhynchus funereus*).

A number of species at Tumut are listed as being rare or threatened or are known to be declining significantly elsewhere in Australia. For example, the Olive Whistler (*Pachycephala olivacea*) and Red-browed Treecreeper (*Climacteris erythrops*) are uncommon species that are of conservation concern in parts of their ranges. Notably, a number of the bird species which are moderately common in the forests at Tumut are known to be declining significantly in the woodlands of south-eastern Australia (Barrett *et al.*, 2003). These include the Crested Shrike Tit (*Falcunculus frontatus*), Rufous Whistler (*Pachycephala rufiventris*), Eastern Yellow Robin (*Eopsaltria australis*) and Scarlet Robin (*Petroica multicolor*) (Reid, 1999; Ford *et al.*, 2001; Cunningham *et al.*, 2008).

As would be expected from such a species-rich assemblage, the birds at Tumut vary substantially in a wide range of life history attributes (Figure 3.6). These include body weight, group type (solitary, pairs or flock), social system (monogamous, polygamous), type of nest (hollow, cup, mud bowl), nest placement (horizontal fork, ground), nesting height, number of eggs laid in a clutch, broods per year and movement behaviour (resident

(a)

(b)

Figure 3.6. A subset of bird species recorded from the Tumut region: (a) Tawny Frogmouth (photo by David Lindenmayer); (b) Eastern Spinebill; (c) White-naped Honeyeater; (d) Grey Fantail (photos by Graeme Chapman).

(c)

(d)

Figure 3.6. (cont.)

versus migrant, latitudinal or altitudinal migrant). They also vary in foraging strategy or guild (*sensu* Mac Nally, 1994) such as:

- **Sweeper** – airborne insects captured on the wing;
- **Hawker** – sally from a perched position to capture airborne insect below the main tree canopy but above bushes; sally to capture airborne insects from the ground or from bushes; gleaning of prey on branches and trunks by perched birds;
- **Pouncer** – capture of prey on the ground from a perched position;
- **Bush carnivore** – gleaning of prey from branches, twigs and tree trunks by perched birds; searching of the ground, often by probing;
- **Ground carnivore** – searching of the ground, often by probing; scratching and searching through the leaf litter;
- **Bark prober** – gleaning of prey on branches and trunks by perched birds; probing and prising bark; searching for prey by tearing bark;
- **Wood searcher** – gleaning of prey on twigs, branches and trunks by perched birds;
- **Foliage searcher** – gleaning of prey on leaves and twigs of trees by perched birds;
- **Nectarivore** – consumption of pollen, nectar and blossoms; gleaning of prey on leaves, branches and trunks of trees by perched birds;
- **Granivore** – consumption of seeds in the understorey; consumption of fallen seeds and seeds of grasses.

The collation of extensive information on the life history attributes of birds has been an important part of the work at Tumut. This is because it has been used to determine whether landscape patterns can be associated with particular groups of species characterised by sets of life history attributes (Lindenmayer *et al.*, 2002a; Chapter 5).

Mammals

There are three broad assemblages of mammals (excluding bats) at Tumut – arboreal marsupials, terrestrial native mammals and terrestrial feral mammals.

Arboreal marsupials

The forests at Tumut support five species of marsupial gliders: the Greater Glider (*Petauroides volans*), Yellow-bellied Glider (*Petaurus australis*), Sugar Glider (*Petaurus breviceps*), Squirrel Glider (*Petaurus norfolcensis*) and

Feathertail Glider (*Acrobates pygmaeus*), and four species of possums: Common Ringtail Possum (*Pseudocheirus peregrinus*), Mountain Brushtail Possum (*Trichosurus cunninghamii*), Common Brushtail Possum (*Trichosurus vulpecula*) and Eastern Pygmy Possum (*Cercartetus nanus*). These species are representatives of five different families and they vary substantially in body size (10–3,500g), breeding system (monogamous to polygamous), diet (specialist exudivores, specialist folivores and omnivores) and home range (<1–85ha) (Lindenmayer, 1997, 2002). Some of these differences are summarised in Table 3.1.

Terrestrial native mammals

The Tumut region supports a rich array of terrestrial native mammals. These include two native murid rodents – the Bush Rat (*Rattus fuscipes*) and Swamp Rat (*Rattus lutreolus*), and two species of small carnivorous marsupials – Agile Antechinus (*Antechinus stuartii*) and Dusky Antechinus (*Antechinus swainsonii*). All of these species are largely absent from Radiata Pine stands but are common in native forests, including patches of remnant native forest surrounded by pine stands (Lindenmayer *et al.*, 1999c). Two relatively rare species in the region include the Spotted-tailed Quoll (*Dasyurus maculatus*) and the Long-nosed Bandicoot (*Parameles nasuta*).

Four species of macropod marsupials commonly observed in the Tumut region are the Swamp or Black Wallaby (*Wallabia bicolor*), Eastern Grey Kangaroo (*Macropus giganteus*), Euro (*Macropus robustus*) and Red-necked Wallaby (*Macropus rufogriseus*). Two other native mammals commonly recorded at Tumut are the Common Wombat (*Vombatus ursinus*) – the world's largest burrowing mammal, and the Echidna (*Tachyglossus aculeatus*) – one of the world's three species of living monotremes. Both occur in a wide range of vegetation types in the study area, including stands of Radiata Pine of a range of age classes (see Table 3.2).

Terrestrial feral mammals

Ten species of terrestrial feral mammals have been recorded in the Tumut region. They are: European Rabbit (*Oryctolagus cuniculus*), Black Rat (*Rattus rattus*), House Mouse (*Mus musculus*), Red Fox (*Vulpes vulpes*), Feral Cat (*Felis catus*), Dingo and/or Wild Dog (*Canis familiaris dingo*), Feral Pig (*Sus scrofa*), Goat (*Capra hircus*), Sambar Deer (*Cervus unicolor*) and Cattle (*Bos taurus*). All are moderately common or abundant, although their distribution and abundance is patchy and is often associated with the

Table 3.1. *Life history attributes of arboreal marsupials recorded in the Tumut region.*

Common name	Home range (ha)	Mean body mass (g)	Social organisation	Mating system[a]	Diet
Feathertail Glider	0.4–2.1	10–14	Colonial (up to 40 individuals); possibly not territorial; up to 29 individuals observed to share a den site	**Promiscuous**; polygynous; monogamous	Arthropods; insect and plant exudates (pollen, honeydew, nectar, seeds)
Sugar Glider	0.5–7.1	115–160 (male) 95–135 (female)	Colonial (up to 12 individuals)	**Polygynous**	Arthropods; insect and plant exudates (wattle gum, eucalypt sap, nectar, pollen, manna, honeydew)
Squirrel Glider	3–5	200–260	Colonial (up to 9 individuals)	**Polygynous**	Arthropods (coleopteran and lepidopteran larvae); insect and plant exudates (wattle gum, eucalypt sap, nectar, pollen)
Yellow-bellied Glider	20–85	450–700	Colonial (up to 11 individuals); up to 5 individuals observed to share a den tree	**Monogamous**; polygynous; polygamous	Arthropods; insect and plant exudates (honeydew, eucalypt sap, nectar, pollen)
Greater Glider	0.7–3.0	900–1,700	Solitary; males' larger home ranges overlap with those of several females; co-occupancy of den trees	**Monogamous**; polygamous; polygynous	Eucalypt foliage specialist; various species of eucalypt taken

Common Ringtail Possum	0.07–2.60	660–900	rare except between matched pairs in the breeding season	**Polygamous**	Foliage of a wide range of plants, flowers, fruits
			Pairs and small groups of animals		
Common Brushtail Possum	0.7–11.0	1,500–3,500	Overlapping male–female pairs, small groups of animals	**Monogamous**; polygamous	Foliage of a wide range of plants, flowers, fruits; occasionally bird nestlings
Mountain Brushtail Possum	1–7	2,500–4,500	Overlapping male–female pairs, small groups of animals	**Monogamous**; polygamous	Foliage of a wide range of plants, seeds, above-ground fungi, truffles
Eastern Pygmy Possum	0.2–1.7	15–40	Small groups of animals	**Polygamous**	Nectar, pollen, fruit; arthropods

[a] The mating system of some (and perhaps many) species of arboreal marsupials is not 'set in concrete'. It may change due to the availability and quality of food resources (Lindenmayer, 2002). The system likely to be most common at Tumut is shown in **bold**.

Table 3.2. *Summary of some life history attributes of terrestrial mammals recorded in the Tumut region.*

Common name	Home range (ha)[a,b]	Mean body mass (g)[a,b]	Social organisation[a,c]	Diet[a,c]
Bush Rat	0.1–1.2[c]	125 (50–225)	Solitary	Arthropods, seeds, fruits, fungi
Swamp Rat	0.2–4.0[a,c]	122 (55–160)	Solitary	Mainly sedges, grasses
Agile Antechinus	0.4–5.3[c]	16–44	Seasonally communal	Invertebrates, small vertebrates
Dusky Antechinus	0.4–1.5[d]	38–170	Solitary	Invertebrates, small vertebrates
Long-nosed Bandicoot	1.5–5.2[e]	850–1,100	Solitary	Invertebrates, seeds, leaves, fungi
Spotted-tailed Quoll	88–2,560[f]	1,500–7,000	Solitary	Small to medium-sized mammals, birds, carrion, invertebrates
Eastern Grey Kangaroo	8–430[c,g]	3,500–66,000	Group	Grasses, forbs
Euro	10–78[g]	6,250–46,500	Group	Grasses, forbs
Red-necked Wallaby	12–32[c]	11,000–27,000	Adult males solitary, females and sub-adults in small groups	Grasses, forbs
Swamp Wallaby	10–27	10,000–20,500	Mainly solitary	Wide variety of plants including fungi, shrubs, grasses, sedges, bracken
Common Wombat	5–23[c]	20,000–39,000 (26,000)	Solitary	Mainly grasses, rushes; also sedges, roots, tubers, mosses, fungi
Echidna	40–70[f]	2,000–7,000	Solitary	Mainly ants and termites; some other invertebrates

(a)

(b)

Figure 3.7. A subset of mammal species recorded from the Tumut region: (a) Sugar Glider (photo by Mike Greer); (b) Greater Glider (photo by David Lindenmayer); (c) Echidna (photo by Esther Beaton); (d) Common Wombat (photo by Esther Beaton).

(c)

(d)

Figure 3.7. (cont.)

extent of human disturbance in different parts of the plantation or adjacent native eucalypt forest. For example, the Black Rat and the House Mouse are primarily associated with stands of recently planted Radiata Pine. Cattle include domesticated animals that are allowed to graze within the boundaries of the plantation as well as animals that have escaped and which roam wild in the neighbouring state forests. The Feral Pig is abundant and hunted heavily, although populations are replenished by hunters who make repeated (illegal) releases of young animals. The Dingo is classified as an introduced species because it arrived in Australia approximately 5,000 years ago (Savolainen et al., 2004), initially as a commensal with indigenous people. Figure 3.7 shows a selection of the mammal species recorded at Tumut.

Reptiles

Reptiles are Australia's most diverse vertebrate group (Wilson and Swan, 2003). However, the reptile assemblage at Tumut is somewhat depauperate relative to many other Australian environments. This is likely to be the result of the predominance of cool and wet montane areas in part of the study area (Jenkins and Bartell, 1980; Green and Osborne, 1994).

Skinks are the most species-rich group at Tumut, and Fischer et al. (2005) recorded 13 species. Two additional species, the Common and Blotched Blue-tongue Lizards (Tiliqua scincoides and T. nigrolutea), are rarely recorded inhabitants of the area. These 15 species vary widely in many aspects of their life history, as shown in Table 3.3.

Elapid snakes are the next most common reptile group encountered at Tumut. They include the Highland Copperhead (Austrelaps ramsayi), Eastern Brown Snake (Pseudonaja textilis), Tiger Snake (Notechis scutatus), Red-bellied Black Snake (Pseudechis porphyriacus) and the White-lipped Snake (Drysdalia coronoides). All of these species are reasonably common and none are regarded as vulnerable or of conservation concern. Figure 3.8 shows a selection of the reptiles species recorded at Tumut.

Surveys of reptiles have employed pitfall traps (Fischer et al., 2005) or have been largely opportunistic, and a number of species present at Tumut may remain undetected. For example, it is likely that the Long-necked Tortoise (Chelodina longicollis) and the Gippsland Water Dragon (Physignathus lesueurii howittii) occur in and around the streamlines, creeks, fire dams and other water bodies around the region.

Frogs

The Tumut region supports at least eight 'species' of frogs (Parris and Lindenmayer, 2004). These are the Common Eastern Froglet (Crinia signifera), Northern Corroboree Frog (Pseudophryne pengilleyi), Brown Brood Frog (Pseudophryne bibronii), Pobblebonk (Limnodynastes dumerilii), Spotted Grass Frog (Limnodynastes tasmaniensis), Smooth Toadlet (Uperoleia laevigata), Emerald Spotted Tree Frog (Littoria peronii) and Whistling Tree Trog (Littoria ewingii) (see Table 3.4 for life history

Table 3.3. *Life history attributes of reptiles recorded during field research at Tumut.*

Common name	Snout–vent length (mm)	Body length (mm)	Life form	Common shelter site	Activity pattern	Mode of thermo-regulation	Mode of reproduction	Clutch/offspring size	Foraging mode	Diet
Eastern Three-lined Skink	60	190	Terrestrial	Debris	Diurnal	Heliotherm	Oviparous	5	Active	Arthropods
Red-throated Skink	70	200	Terrestrial	Leaf litter	Diurnal	Heliotherm	Oviparous	7	Active	Arthropods
Black Rock Skink	110	210	Saxicolous	Crevices	Diurnal	Heliotherm	Viviparous	4	Sit and wait	Insects
White's Skink	75	180	Terrestrial	Rocks	Diurnal	Heliotherm	Viviparous	4	Active	Arthropods
Southern Water Skink	75	230	Terrestrial	Debris	Diurnal	Heliotherm	Viviparous	4	Active	Arthropods
Three-toed Skink	50	150	Fossorial	Debris	Diurnal	Thigmotherm	Viviparous	3	Active	Ants/Termites
Grass Skink	40	110	Terrestrial	Leaf litter	Diurnal	Heliotherm	Oviparous	3	Active	Arthropods
Garden Skink	40	110	Terrestrial	Leaf litter	Diurnal	Heliotherm	Oviparous	3	Active	Arthropods
Highlands Forest Skink	45	120	Fossorial	Leaf litter	Nocturnal	Thigmotherm	Oviparous	4	Active	Insects
Coventry's Skink	45	120	Terrestrial	Debris	Diurnal	Heliotherm	Viviparous	5	Active	Arthropods
Southern Grass (Mt. Log) Skink	50	130	Terrestrial	Leaf litter	Diurnal	Heliotherm	Viviparous	5	Active	Insects
Spencer's Skink	50	130	Terrestrial	Crevices	Diurnal	Heliotherm	Viviparous	3	Active	Insects
	75	230	Terrestrial	Debris	Diurnal	Heliotherm	Viviparous	4	Active	Arthropods

Species										
Yellow-bellied Water Skink										
Blotched Blue-tongue	250	370	Terrestrial	Debris	Diurnal	Heliotherm	Viviparous	10	Active	Invertebrates
Common Blue-tongue	260	440	Terrestrial	Debris	Diurnal	Heliotherm	Viviparous	15	Active	Invertebrates
Jacky Lizard	90	290	Semi-arboreal	Debris	Diurnal	Heliotherm	Oviparous	10	Sit and wait	Insects
Eastern Stone Gecko	50	110	Terrestrial	Rocks	Nocturnal	Thigmotherm	Oviparous	2	Active	Arthropods
Tree (Crevice) Skink	100	200	Arboreal/ Saxicolous	Crevices	Diurnal	Heliotherm	Viviparous	4	Sit and wait	Insects
Olive Legless Lizard	100	300	Terrestrial	Debris	Diurnal/ Nocturnal	Thigmo/ heliotherm	Oviparous	2	Active	Insects
Southern Rainbow Skink	45	120	Terrestrial	Leaf litter	Diurnal	Heliotherm	Oviparous	2	Active	Ants/ Termites
Boulenger's Skink	45	120	Terrestrial	Debris	Diurnal	Heliotherm	Oviparous	4	Active	Arthropods
Eastern Striped Skink	80	250	Terrestrial	Rocks	Diurnal	Heliotherm	Oviparous	5	Active	Arthropods
Copper-tailed Skink	60	200	Terrestrial	Rocks	Diurnal	Heliotherm	Oviparous	5	Active	Arthropods
Marbled Gecko	45	100	Arboreal/ Saxicolous	Bark	Nocturnal	Thigmotherm	Oviparous	2	Active	Insects

(a)

(b)

Figure 3.8. A subset of the reptile species recorded from the Tumut region: (a) Three-toed Skink (photo by Damian Michael); (b) Copperhead Snake; (c) Blue-tongue Lizard; (d) Tiger Snake (photos by David Lindenmayer).

(c)

(d)

Figure 3.8. (cont.)

attributes of these frogs). *Littoria ewingii* is not a single taxon but rather a complex of closely related species for which the taxonomy has yet to be fully resolved. Of the species recorded from Tumut, only the Northern Corroboree Frog is of conservation concern. It is listed as vulnerable in

Table 3.4. *Life history attributes of frogs recorded during field research at Tumut.*

Common name	Scientific name	Body size (mm)	Habitat and breeding	No. of eggs	Tadpole type	Tadpole description
Common Eastern Froglet	*Crinia signifera*	18–25 (♂) 19–28 (♀)	Extremely widespread and highly variable species found in a variety of habitats including those heavily modified by humans. Eggs are laid in clumps attached to sticks, grass or leaf litter in shallow water	100–150	Lentic	As in the case of adults, tadpoles can be extremely variable
Northern Corroboree Frog	*Pseudophryne pengilleyi*	26–28 (♂) 26–30 (♀)	Typically occurs in high-elevation bogs and swamps. Eggs are laid in burrows under moss, grass and sedges	10–30	Lentic	Dark brown tadpoles with striking black and yellow colour pattern developing as metamorphosis approaches
Brown Brood Frog	*Pseudophryne bibronii*	22–30 (♂) 25–32 (♀)	Occurs in areas with damp leaf litter and soil. Eggs are laid under logs, stones or leaf litter	75–190	Lentic	Dark brown tadpoles with mottled tail fin reaching 31 mm
Pobblebonk	*Limnodynastes dumerilii*	52–70 (♂) 52–83 (♀)	Breeds in a wide variety of habitats including permanent and temporary waterholes. Floating foam nests made in aquatic areas	3,900	Lentic	Dark-grey to black tadpoles
Spotted Grass Frog	*Limnodynastes tasmaniensis*	31–42 (♂) 32–47 (♀)	Breeds in a wide variety of marshy habitats associated with permanent or temporary waterholes. Foam nests are attached to aquatic vegetation	85–1,360	Lentic	Dark-brown to black tadpoles reaching 46 mm

Smooth Toadlet	*Uperoleia laevigata*	20–28 (♂) 22–32 (♀)	Breeds in a variety of areas including temporarily flooded grasslands	–	Lentic	
Emerald Spotted Tree Frog	*Littoria peronii*	44–53 (♂) 46–55 (♀)	Breeds in a wide variety of habitats including permanent and temporary waterholes	–	Lentic	Pale golden-yellow tadpoles up to 44mm
Whistling Tree Frog	*Littoria ewingii*	20–40 (♂) 32–46 (♀)	Breeds in a wide variety of habitats including permanent and temporary waterholes Eggs attached to clumps of submerged vegetation	–	Lentic	Transparent tadpoles up to 50 mm

Based on information in Barker *et al.* (1995) and Tyler (1999).

(a)

(b)

Figure 3.9. A subset of the frog species recorded from the Tumut region: (a) Peron's Tree Frog (photo by Debbie Claridge); (b) Eastern Banjo Frog (photo by Esther Beaton); (c) Northern Corroboree Frog (photo by David Hunter); (d) Spotted Marsh Frog (photo by LiquidGhoul: GNU Free Documentation Licence).

(c)

(d)

Figure 3.9. (cont.)

New South Wales and nationally (Department of Environment, Heritage and the Arts, 2008). Figure 3.9 shows some of the frog species recorded at Tumut.

Other groups

No-one to date has completed a systematic survey of the freshwater fish inhabiting the region. Similarly, there has been no detailed work on microchiropteran bats, but opportunistic observations completed during other surveys suggest that 15 or more species may occur in the area.

The invertebrate fauna of the Tumut is diverse and species-rich, although only small-scale sampling has been conducted to date (e.g. Schmuki *et al.*, 2008). Work by Smith (2006) found almost 70 species of ants, over 105 species of flies and nearly 80 species of beetles. However, sampling was restricted to a limited number of sites and a limited period, and also to one survey methodology (pitfall trapping), suggesting that a large number of additional species occur in the area.

Summary

The Tumut region is dominated by a large industrial Radiata Pine plantation. An array of patches of remnant native forest occur within the boundaries of the plantation. These patches vary in size, shape, vegetation type and many other features. The same vegetation types occur in a contiguous native forest setting in the Bondo State Forest and the Brindabella and Kosciuszko National Parks, which are adjacent to the plantation. Such similarities are critical to the design of the major cross-sectional study at Tumut – which is the topic of the next chapter.

4 · Setting up the study: the design and implementation of the main cross-sectional study at Tumut

This chapter focuses on the major cross-sectional study conducted at Tumut and the many steps associated with its design and implementation. It also describes why the Tumut area was chosen for study. The cross-sectional study provides a platform for a raft of other linked and/or subsidiary projects.

The experimental design underpinning the cross-sectional study at Tumut

The experimental design underpinning the main cross-sectional or comparative study at Tumut entailed a series of logically linked and often iterative steps (Figure 4.1). The process of experimental design involved extensive dialogue between a team of highly qualified professional statisticians and field ecologists. Particular ecological questions were posed and then a set of possible experimental designs developed to answer them. Different designs would have entailed different intensities of field survey. There was also the issue of the amount of survey required within different-sized patches of remnant native eucalypt forest within the plantation to provide an adequate estimate of animal presence and abundance. These considerations meant that the final experimental design was preceded by an iterative process of defining and redefining ecological questions, considering different experimental designs and completing pilot field surveys, survey feasibility and preliminary data analysis.

For the purposes of the major natural experiment at Tumut, vegetation cover was assigned into three broad landscape components termed landscape context classes. They were:

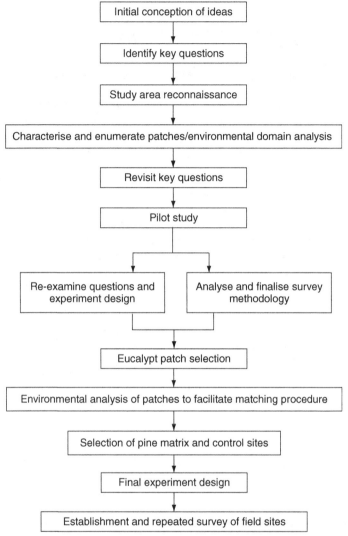

Figure 4.1. Sequence of steps involved in the experimental design of the main study at Tumut.

- Landscape Context Class 1: Remnant patches of native eucalypt forest (see Figure 3.4 in Chapter 3). These eucalypt remnants vary in size (1–124 ha), shape (long and narrow versus elliptical or round), vegetation type and other features.
- Landscape Context Class 2: Extensive areas of exotic Radiata Pine forest which surrounds the eucalypt remnants. These parts of the

landscape are sometimes referred to as the 'landscape matrix' in this book.

- Landscape Context Class 3: Large continuous areas of native eucalypt forest that occur beyond the boundaries of the plantation and which include the Kosciusko and Brindabella National Parks and the Bondo and Bungongo State Forests.

The major investigation at Tumut was designed as a comparative study. That is, landscape context effects on biota in three broad landscape context classes were compared: sites in eucalypt remnants surrounded by extensive stands of Radiata Pine (a group of sites termed Landscape Context Class 1); sites dominated by Radiata Pine trees that surrounded the eucalypt remnants and which formed the landscape matrix in the study region (Landscape Context Class 2); and sites located within large continuous areas of native eucalypt forest (Landscape Context Class 3).

Enumeration of the eucalypt remnants

The initial stage of the cross-sectional study was to enumerate the eucalypt remnants that occurred in our study area. Data on the size and spatial location of remnants embedded within the Radiata Pine plantation were extracted from a geographical information system (GIS) (ARCINFO, Esri, CA) developed by State Forests of NSW. A total of 192 remnants of remnant *Eucalyptus* spp. forest, of varying size, shape and vegetation

---

Box 4.1. A different design for a different study

The aim of the cross-sectional study at Tumut was to quantify the effects of landscape context. The design would have been different if the principal objective had been to estimate animal abundance throughout a patchy and extensively modified forest. For such a study, the interest would have been in obtaining a precise (minimum variance) unbiased estimate of the overall density of different biotic groups in the landscape. In that case, sampling intensity may have depended on the relative area of forest in each spatial context class (i.e. remnants and sites dominated by stands of Radiata Pine and contiguous eucalypt forest). The design methodology required to address a problem of that nature is generally known as a survey sampling design.

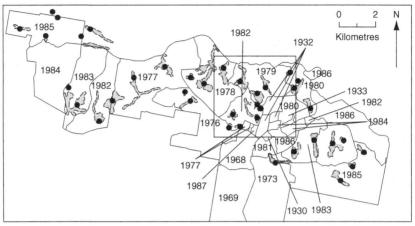

Figure 4.2. Part of the pilot study area at Tumut showing the time of establishment of pine stands and hence the time since isolation of eucalypt remnants (shaded).

type, had escaped clearing for plantation establishment. The eucalypt remnants varied from 0.1 to 124ha in size and were distributed widely throughout the plantation (Figure 4.2). Other data such as the shape of each patch and the age of the surrounding forest were also extracted from the GIS database.

After interrogating the GIS database and enumerating the eucalypt remnants at Tumut, each remnant was walked and additional data were gathered, including dominant tree species, topographic position (i.e. gully, mid slope, ridge or flat), slope of the site, geology and the characteristics of the site (rocky area, steep hilltop, swamp or streamside reserve). Table 4.1 is a cross-tabulation of the characteristics of the 192 remnants. This shows that there were many different reasons why the various remnants were retained, and this meant that there was ample opportunity for considerable representation and replication of classes. This provided the basis for obtaining a stratified random sample of sites (see below).

Selection procedures for eucalypt remnants surrounded by stands of Radiata Pine

Of the 192 remnants scattered throughout the Radiata Pine plantation, 41 measured >1 ha in size and were too small to enable the sampling effort to be consistent with other broad types of sites. As a result, these sites were assigned to a separate micro-patch study (see below) and deleted from the pool of remnants available for selection. A sample of 86 remnants was then

Table 4.1. *Cross-tabulation of patch features in the initial enumeration phases of the study. The characteristics of 192 remnants of eucalypt forest in different shape and size classes are classified by: (a) the nature of the surrounding vegetation; (b) dominant tree species; (c) topographic position; (d) reasons for retention and (e) the age of the surrounding vegetation.*

(a)

Remnant shape	Size (ha)	Nature of surrounding vegetation	
		Uniform	Heterogeneous
Patch			
	<0.8	16	2
	0.8–3	13	6
	3–10	13	6
	11–20	6	11
	>20	5	4
Strip			
	<0.8	3	1
	0.8–3	9	2
	3–10	27	10
	11–20	15	7
	>20	21	6

(b)

Remnant shape	Size (ha)	Dominant tree species in remnant[a]				
		Cm	Vi	Ra	Dry	Other
Patch						
	<0.8	1	4	8	4	1
	0.8–3	4	1	11	12	0
	3–10	2	4	8	3	2
	11–20	3	5	6	3	0
	>20	0	2	4	2	1
Strip						
	<0.8	1	1	2	0	0
	0.8–3	6	3	1	1	0
	3–10	15	14	4	3	1
	11–20	7	5	4	4	2
	>20	11	6	3	1	6

[a] The codes for the tree species are: **Cm** (*Eucalyptus camphora*), **Vi** (*E. viminalis*), **Ra** (*E. radiata*), **Dry** (*E. dives, E. macrorhynca* and *E. bridgesdiana*) and **Other** (*E. dalrympleana, E. pauciflora* and *E. stellulata*).

Table 4.1. *(cont.)*

(c)

Remnant shape	Size (ha)	Topographic position	
		Gully	Mid slope, ridge, flat
Patch			
	<0.8	5	13
	0.8–3	6	22
	3–10	2	17
	11–20	4	13
	>20	4	5
Strip			
	<0.8	2	2
	0.8–3	9	2
	3–10	33	4
	11–20	17	5
	>20	26	1

(d)

Remnant shape	Size (ha)	Reasons for retention	
		Water values	Other reasons
Patch			
	<0.8	2	16
	0.8–3	3	25
	3–10	1	18
	11–20	2	15
	>20	1	8
Strip			
	<0.8	0	4
	0.8–3	9	2
	3–10	33	4
	11–20	14	8
	>20	23	4

(e)

Remnant shape	Size (ha)	Age of the surrounding vegetation	
		Young	Old
Patch			
	<0.8	13	5
	0.8–3	25	3

Table 4.1 (*cont.*)

Remnant shape	Size (ha)	Age of the surrounding vegetation	
		Young	Old
	3–10	14	5
	11–20	6	11
	>20	5	4
Strip			
	<0.8	2	2
	0.8–3	6	5
	3–10	22	15
	11–20	14	8
	>20	11	16

selected from the remaining 141 remnants using a stratified random and replicated statistical procedure. The strata for selection were:

- Four patch size classes (1–3 ha, 4–10 ha, 11–20 and >20 ha).
- Two isolation age classes: less than and greater than 20 years since the surrounding landscape had been converted to pine stands. The division for these isolation times was arbitrary but distinguished between patches created during two major periods of plantation expansion, when most of the original eucalypt forest was cleared.
- Five dominant eucalypt forest type classes:
- Ribbon Gum
- Narrow-leaved Peppermint
- Mountain Swamp Gum
- 'Dry forest': Red Stringybark and Apple Box
- 'Other forest': Mountain Gum, Snow Gum and Black Sallee.

As many combinations of the major stratifying variables were chosen as possible. Replicates of these combinations also were selected where available. For example, the four classes of remnant size contained 20 sites (1–3 ha), 21 sites (3–10 ha), 21 sites (11–20 ha) and 24 sites (>20 ha). Similarly, the five forest type classes contained 14 sites ('Dry forest'), 21 sites (Narrow-leaved Peppermint forest), 19 sites (Ribbon Gum forest), 13 sites ('Other forest') and 19 sites (Swamp Gum forest).

Where possible, eucalypt remnants were located at least 1 km apart to reduce the potential for spatial dependence in the field data. The suite of remnants selected, and their associated design features, are set out in Table 4.2.

Table 4.2. *Selection of eucalypt remnants at Tumut and their stratifying design characteristics.*

Site	Size class (ha)	Surrounding forest age	Surrounding forest type	Shape	Tree species[a]
31	1–3	Young	Uniform	Patch	Dry
33	1–3	Young	Uniform	Patch	Ra
37	3–10	Young	Uniform	Patch	Ra
56	>20	Young	Uniform	Strip	Dry
198	>20	Old	Uniform	Strip	Cm
255	11–20	Young	Uniform	Patch	Dry
310	11–20	Old	Uniform	Strip	Vi
335	1–3	Old	Heterogeneous	Patch	Ra
369	11–20	Young	Heterogeneous	Strip	Ra
418	1–3	Old	Uniform	Patch	Cm
422	>20	Young	Heterogeneous	Strip	Cm
449	3–10	Young	Uniform	Strip	Vi
490	1–3	Young	Uniform	Patch	Ra
508	1–3	Young	Uniform	Patch	Vi
559	1–3	Young	Uniform	Strip	Ra
569	1–3	Young	Uniform	Strip	Dry
571	>20	Old	Heterogeneous	Patch	Dry
632	3–10	Young	Uniform	Patch	Vi
657	3–10	Young	Uniform	Strip	Cm
740	3–10	Young	Heterogeneous	Patch	Ra
801	11–20	Young	Heterogeneous	Strip	Cm
802	3–10	Old	Uniform	Strip	Vi
810	1–3	Young	Uniform	Patch	Dry
819	1–3	Young	Heterogeneous	Patch	Dry
828	3–10	Young	Heterogeneous	Patch	Ra
829	3–10	Young	Uniform	Patch	Dry
838	11–20	Young	Heterogeneous	Patch	Cm
872	1–3	Old	Uniform	Patch	Dry
876	11–20	Old	Uniform	Patch	Dry
897	>20	Young	Heterogeneous	Patch	Vi
900	1–3	Young	Uniform	Patch	Cm
906	11–20	Old	Uniform	Strip	Cm
917	3–10	Old	Uniform	Patch	Vi
1129	1–3	Old	Uniform	Strip	Cm
1186	11–20	Old	Heterogeneous	Patch	Ra
1263	3–10	Old	Uniform	Patch	Cm
1364	3–10	Old	Uniform	Strip	Cm
1401	>20	Old	Uniform	Strip	Vi
1434	3–10	Old	Heterogeneous	Patch	Dry
1495	>20	Old	Uniform	Patch	Other

Table 4.2. (*cont.*)

Site	Size class (ha)	Surrounding forest age	Surrounding forest type	Shape	Tree species[a]
1537	11–20	Old	Uniform	Patch	Ra
1630	11–20	Young	Heterogeneous	Patch	Ra
1816	>20	Old	Uniform	Strip	Other
1837	11–20	Old	Uniform	Patch	Vi
1854	>20	Young	Heterogeneous	Strip	Other
1863	11–20	Old	Uniform	Strip	Other
1875	1–3	Old	Uniform	Strip	Vi
1902	11–20	Old	Heterogeneous	Patch	Vi
1908	11–20	Old	Heterogeneous	Patch	Ra
1916	3–10	Old	Heterogeneous	Strip	Other
276b	11–20	Young	Heterogeneous	Patch	Vi
A4	>20	Old	Uniform	Patch	Ra
A5	3–10	Young	Uniform	Patch	Ra
A6	>20	Young	Uniform	Patch	Ra
B2	3–10	Young	Heterogeneous	Strip	Ra
B5	1–3	Young	Heterogeneous	Strip	Vi
B6	>20	Young	Uniform	Strip	Other
C	>20	Old	Uniform	Strip	Ra
C3	11–20	Young	Uniform	Patch	Ra
C5	3–10	Young	Uniform	Patch	Vi
D1	1–3	Old	Uniform	Strip	Cm
D3	3–10	Young	Uniform	Strip	Vi
E1	3–10	Young	Uniform	Patch	Cm
E3	11–20	Young	Uniform	Strip	Vi
F1	>20	Young	Uniform	Patch	Vi
F2	>20	Old	Heterogenous	Strip	Cm
F3	11–20	Young	Uniform	Strip	Other
F4	3–10	Young	Uniform	Patch	Ra
G3	>20	Old	Uniform	Strip	Other
H1	>20	Young	Uniform	Strip	Cm
I1	3–10	Young	Uniform	Strip	Cm
I2	3–10	Young	Uniform	Strip	Dry
I4	1–3	Young	Uniform	Strip	Dry
J3	>20	Old	Uniform	Strip	Vi
K1	3–10	Young	Uniform	Strip	Ra
K3	>20	Old	Uniform	Strip	Cm
L3	1–3	Young	Heterogenous	Patch	Ra
O	1–3	Young	Uniform	Strip	Cm
O3	3–10	Young	Heterogenous	Strip	Cm
P	3–10	Young	Uniform	Patch	Other
P3	11–20	Old	Uniform	Strip	Dry

Table 4.2. (*cont.*)

Site	Size class (ha)	Surrounding forest age	Surrounding forest type	Shape	Tree species[a]
T	11–20	Young	Uniform	Strip	Ra
U3	>20	Young	Uniform	Strip	Vi
Y3	>20	Young	Uniform	Strip	Ra
Z	3–10	Young	Uniform	Strip	Dry

[a] The codes for the tree species are: **Cm** (*Eucalyptus camphora*), **Vi** (*E. viminalis*), **Ra** (*E. radiata*), **Dry** (*E. dives, E. macrorhynca* and *E. bridgesdiana*) and **Other** (*E. dalrympleana, E. pauciflora* and *E. stellulata*).

In the case of the selection of areas of remnant eucalypt forest, the experimental unit was a remnant and not observations within a remnant. In comparative studies, there are compelling statistical arguments in favour of increasing the number of experimental units rather than increasing the number of observations per unit. Accordingly, 86 remnants were selected, rather than the alternative of choosing multiple sites in fewer remnants.

Identifying a sampling unit: a pilot study and calibration of survey methods

As outlined above, the main cross-sectional study at Tumut was a comparative investigation of biota in sites within different landscape context classes. It was not appropriate to sample remnants proportional to their area. Moreover, this would have been nonsensical in the 50,000 ha of pine stands that comprised the landscape matrix surrounding the eucalypt remnants. Nor would have it been sensible in the 100,000+ha of contiguous native eucalypt forests that contained the matched control areas for the study. In the specific case of the eucalypt remnants that varied by an order of magnitude in size, complete patch searches would have confounded patch size with sampling intensity – a not uncommon problem in many studies in human-modified landscapes. Thus, for this comparative investigation, sampling intensity for different eucalypt remnants needed to be independent of remnant size. Sampling intensity also needed to be equivalent across eucalypt remnants, eucalypt 'controls' and for sites dominated by stands of Radiata Pine.

A key issue then became: 'What is an appropriate experimental unit of fixed size?' That is, what length of transect within a site would give a reasonable (normalised) estimate of animal presence and abundance per unit area for a eucalypt remnant, for a eucalypt control and for a Radiata Pine site? A pilot study was needed to tackle this problem. Arboreal marsupials were used as the target group in the pilot study.

A 5,500 ha area in the northern part of the Buccleuch State Forest was targeted for the pilot study (Figure 4.2). A total of 40 eucalypt remnants were selected for survey.

Each remnant was surveyed twice for arboreal marsupials in the pilot study: once by a complete patch search, and once by spotlighting a randomly chosen 600 m transect within each remnant. Two experienced observers were responsible for the spotlighting. These surveys were designed to determine whether there were systematic differences in the mean density of animals between observers or counting methods (i.e. a complete patch search or a transect).

Spotlighting surveys were completed during September and October 1995 and were confined to clear, still and dry nights, since environmental conditions such as temperature, rainfall and high winds can influence the detectability of arboreal marsupials (Davey, 1990; Laurance, 1990). In addition, surveys were limited to 3–4 hours' duration on any given night. This was for two reasons:

(1) Observer fatigue after a prolonged period of spotlighting could have resulted in animals being missed;
(2) Arboreal marsupials appear to be most active after emergence and their activity patterns decline with increasing time after dusk (Thomson and Owen, 1964).

Thus, we attempted to limit our surveys to those times when the animals were likely to be most active.

A series of lines, each 50 m apart and marked with coloured flagging tape, were set out on each of the 40 sites in the pilot study area prior to the commencement of spotlighting. The distance along each transect was recorded using a hip-chain. Flagging tape of different colours was used to delimit different 200 m sections of a given line. These procedures:

• Ensured that the entire area of eucalypt forest within each of the 40 remnants could be surveyed;
• Allowed a 100–600 m-long transect to be established for comparison with the results of complete patch searches;
• Enabled the exact location of animals to be recorded.

The experimental design for the pilot study was as follows. The 40 remnants in the pilot study area represented four size classes; 1–3 ha, 3–10 ha, 11–20 ha and >20 ha. Sites were grouped into 20 'like' pairs that were similar in size. Two observers were then assigned at random to remnants within these pairs, giving a total of 20 sites for each person to survey. A total of five sets of sites, with each set comprising four sites, was constructed. These sets contained one representative of each of the four size classes. Each observer completed spotlighting of the nominated set of four sites before moving to the next set of remnants. For each set, an observer was randomly assigned a given survey method (i.e. complete site search or 100–600 m transect). The alternative counting method was employed when the sets were spotlighted the second time. The design used to allocate observers to sites and methods is given in Table 4.3.

For each of the 40 remnants in the pilot study, we recorded data on:

- The presence and abundance of the different species of arboreal marsupials;
- Animal location along a given transect;
- Attributes of the sites including topographic position (i.e. gully, mid slope, ridge or flat), dominant tree species, slope of the site, remnant shape, geology, the type of site (rocky area, steep hilltop, swamp, streamside reserve) and the age of the surrounding Radiata Pine forest.

The 40 remnants targeted for survey in the pilot study totalled 436.7 ha of eucalypt forest. These sites varied in size from 0.4 to 40.4 ha in size and had been surrounded by Radiata Pine stands for 15–40 years (Figure 4.2). The dominant species of tree in the remnants were Narrow-leaved Peppermint, Swamp Gum, Red Stringybark and Ribbon Gum.

The complete patch searches of the 40 remnants in the pilot study resulted in 69.9 km of transects being spotlighted. The total spotlighting time in this case was approximately 65 hours. The sample transects (200–600 m in length) summed to 17.6 km, which corresponded to approximately 25 hours of spotlighting across all 40 sites. Data on arboreal marsupials that were generated from these surveys are presented in Table 4.4. For both counting methods, count data were normalised so that the response variable was the abundance of animals per 3 ha.

Analysis of variance of the mean density of animals showed no significant difference between estimates obtained by the complete patch search and a randomly selected 600 m site. In addition, there was no significant difference between the estimates of abundance recorded by different observers.

Table 4.3. *The design used to allocate observers to different sets of patches as part of a pilot study to examine differences in the ability to detect arboreal marsupials. The four size classes were:* **Class 1**, *1–3 ha;* **Class 2**, *3–10ha;* **Class 3**, *11–20ha;* **Class 4**, *>20ha. Sites were allocated to 20 'like' pairs of roughly similar size and the identity of the matched pair is given in Column 5.*

Observer	Survey technique	Site identity	Size class	Pair
1	Complete survey	258	2	10
1	Complete survey	353	3	15
1	Complete survey	B3	4	20
1	Complete survey	T3	1	5
2	Complete survey	S3	1	2
2	Complete survey	I4	3	12
2	Complete survey	661	4	17
2	Complete survey	276a	2	7
1	Site search	422	1	2
1	Site search	369	3	12
1	Site search	102	2	7
1	Site search	700	4	17
2	Site search	E3	1	5
2	Site search	567	4	20
2	Site search	335	3	15
2	Site search	C3	2	10
1	Complete survey	446	2	8
1	Complete survey	310	1	3
1	Complete survey	632	3	13
1	Complete survey	569	4	18
2	Complete survey	235	2	9
2	Complete survey	115	3	14
2	Complete survey	325	4	19
2	Complete survey	I1	1	4
1	Site search	114	4	16
1	Site search	433	1	1
1	Site search	449	3	11
1	Site search	V3	2	6
2	Site search	198	1	1
2	Site search	I2	3	11
2	Site search	389	4	16
2	Site search	D3	2	6
1	Complete survey	418	2	9
1	Complete survey	490	4	19
1	Complete survey	K1	3	14
1	Complete survey	U3	1	4
2	Complete survey	164	1	3

Table 4.3. (*cont.*)

Observer	Survey technique	Site identity	Size class	Pair
2	Complete survey	599	2	8
2	Complete survey	276b	3	13
2	Complete survey	365	4	18
1	Site search	T3	1	5
1	Site search	353	3	15
1	Site search	B3	4	20
1	Site search	258	2	10
2	Site search	276a	2	7
2	Site search	661	4	17
2	Site search	I4	3	12
2	Site search	S3	1	2
1	Complete survey	700	4	17
1	Complete survey	102	2	7
1	Complete survey	422	1	2
1	Complete survey	369	3	12
2	Complete survey	C3	2	10
2	Complete survey	335	3	15
2	Complete survey	E3	1	5
2	Complete survey	567	4	20
1	Site search	569	4	18
1	Site search	632	3	13
1	Site search	446	2	8
1	Site search	310	1	3
2	Site search	325	4	19
2	Site search	I1	1	4
2	Site search	235	2	9
2	Site search	115	3	14
1	Complete survey	114	4	16
1	Complete survey	433	1	1
1	Complete survey	449	3	11
1	Complete survey	V3	2	6
2	Complete survey	198	1	1
2	Complete survey	I2	3	11
2	Complete survey	D3	2	6
2	Complete survey	389	4	16
1	Site search	U3	1	4
1	Site search	K1	3	14
1	Site search	418	2	9
1	Site search	490	4	19
2	Site search	164	1	3
2	Site search	599	2	8
2	Site search	276b	3	13
2	Site search	365	4	18

Table 4.4. *Count data for arboreal marsupials in the pilot study.*

Species	Complete patch survey		Site search	
	No. animals	No. remnants	No. animals	No. remnants
Greater Glider (*Petauroides volans*)	80	20	35	13
Common Ringtail Possum (*Pseudocheirus peregrinus*)	84	23	34	16
Sugar Glider (*Petaurus breviceps*)	6	5	0	0
Feathertail Glider (*Acrobates pygmaeus*)	1	1	0	0
Mountain Brushtail Possum (*Trichosurus caninus*)	16	11	3	2
Common Brushtail Possum (*Trichosurus vulpecula*)	23	7	1	1

Graphs showing the relationship between the sample count and the complete patch search are shown in Figure 4.3 for the total number of animals and the abundance of the Greater Glider and the Common Ringtail Possum. Data were too sparse to show relationships for other species of arboreal marsupials. Estimates and associated 95% confidence intervals for the slopes of the regression lines for these three relationships (on a log–log scale) were: 0.79 (0.47, 1.11), 1.33 (0.90, 1.75) and 0.95 (0.59, 1.31), respectively. All statistics were obtained by Poisson regression analysis (McCullagh and Nelder, 1989). The 95% confidence intervals (given in parenthesis above) showed that the slopes were not significantly different from 1 (see examples in Figure 4.3). These results confirmed that there was no bias from using a single random 600m transect to estimate the density of arboreal marsupials within remnant patches.

Poisson regression analysis was also used to check the veracity of significant covariates for use as stratifying variables for the design of the main study. In this case, the response variable was the density of animals per 3 ha averaged from the complete patch search and the transect. Significant covariates were identified for three responses. These were the total abundance of arboreal marsupials, the abundance of the Greater Glider and the abundance of the Common Ringtail Possum. The significant explanatory variables in the models developed were remnant size, remnant shape, dominant tree species and the age and heterogeneity of the surrounding Radiata Pine plantation. These results confirmed the validity of

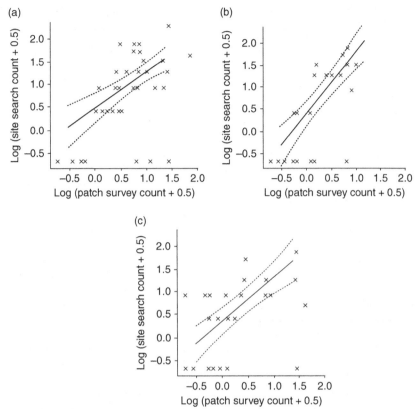

Figure 4.3. Examples of relationships between whole patch counts and transect surveys for arboreal marsupials in the pilot study.

these five variables as major stratifying variables in the subsequent main study (see above).

In summary, the pilot study provided a robust statistical basis for the identification of a common experimental unit (a 600 m-long transect) that was independent of remnant size, which was appropriate for our comparative investigation and ensured that estimates of animal occurrence in eucalypt remnants was not confounded with sampling intensity. Data from the pilot study also confirmed the value of the stratifying variables used in the selection of the 86 remnants in the experimental design.

Site establishment for eucalypt remnants

Based on the outcomes of the pilot study, a permanent transect was established in each of the 86 eucalypt remnants that were identified through the randomised and replicated patch selection procedure. In the 63 eucalypt remnants that were 3 ha or larger in size, a 600 m long × 50 m wide transect was established. Thus, the area of forest surveyed on these sites summed to 3 ha. The length of transects in the remaining 23 remnants, which ranged from 1 to 3 ha in size, was scaled according to the area of the site: a 200 m transect was set out in sites 1–2 ha in size and one of 400 m in length was used in remnants measuring between 2 and 3 ha. Transects in all eucalypt remnants typically started at an edge and were directed towards its centre; the starting point was chosen at random.

Selection of eucalypt 'controls' and Radiata Pine 'matrix' sites

In addition to the 86 remnant eucalypt sites, sites also were selected in Radiata Pine stands as well as sites in large areas of contiguous native *Eucalyptus* forest adjacent to the plantation. As in the case of eucalypt remnants 3 ha or larger, a permanent 600 × 50 m transect was established. Thus, the area of forest surveyed on these sites summed to 3 ha.

Sites in the eucalypt controls and in the Radiata Pine matrix were matched with the eucalypt remnants on the basis of topographic position, climate, geology and forest type (except the Radiata Pine sites). The matching procedure was facilitated by extensive field reconnaissance as well as careful inspection of climatic and regional environmental datasets which included:

- Digital maps of coarsely classified vegetation data developed from previous state government-funded field surveys in the region. These highlighted the locations of broad types of native vegetation (e.g. moist forest, dry forest, woodland, exotic plantation).
- Contour information digitised from 1:25,000 topographic scale maps and stored in a GIS database to highlight variation across the Tumut region for important terrain features such as slope.
- Surfaces of estimated climatic conditions generated using the computer-based climate analysis package BIOCLIM (Nix and Switzer, 1991).

The 40 sites in large areas of contiguous *Eucalyptus* forest included eight replicates of each of the five forest types used in the selection of remnant areas within the Radiata Pine plantation. The sites were positioned around the southern, eastern and northern boundaries of the plantation to reduce potential problems of location-specific phenomena or geographical bias influencing the results – such as a past event (e.g. a major wildfire) having eliminated all the animals from a given area. Finally, the sites in contiguous eucalypt forest were located at least 1 km (and usually more than 2 km) from other sites, so as to limit potential problems associated with spatial dependence in the data (see above).

A total of 40 sites was selected that were dominated by stands of Radiata Pine. These sites were scattered throughout the plantation to reduce the influence of geographical bias and spatial dependence on the results. Environmental and climatic datasets were examined to help match the Radiata Pine sites to the remnants and sites in contiguous eucalypt forest. Studies of the vertebrate fauna in Radiata Pine forest by other workers have indicated that the age of the forest and its stage of development in the silvicultural cycle (e.g. unthinned versus first and second thinning) may influence the abundance of animals (Suckling *et al.*, 1976; Friend, 1980; Rishworth *et al.*, 1995a, 1995b). In light of this, Radiata Pine-dominated sites were distributed among different age classes in the plantation.

Measurement of covariates

In addition to the major design variables in the study, a range of other variables were estimated for, or measured at, each site. These covariates were classified into three broad categories:

- Climatic parameters;
- Landscape location and spatial context;
- Vegetation structure and plant species composition.

The climate analysis program BIOCLIM (Nix, 1986) was employed to generate estimates of a wide range of bioclimatic variables for each site. Plant-growth response estimates were derived using the computer-based program GROCLIM. GROCLIM is an extension of the program GROWEST, which is a simple model of the response of plants to key primary environmental characteristics such as light, heat and water regimes.

The list of parameters derived from the use of BIOCLIM and GROCLIM is presented in Table 4.5. Many of these variables were highly autocorrelated. To avoid statistical problems due to collinearity which arise from this constraint, principal components analysis (Digby and Kempton, 1987) was used to calculate the first two principal components of the original attributes. This enabled orthogonalisation and reduction of the number of climate variables, to facilitate subsequent statistical analyses.

Table 4.5. *BIOCLIM and GROCLIM variables used in climatic characterisation of sites.*

BIOCLIM and GROCLIM variables

1.	Annual mean temperature (°C)
2.	Mean diurnal range/mean (monthly max–min)
3.	Isothermality
4.	Temperature seasonality (CV – coefficent of variation)
5.	Maximum temperature of the hottest month (°C)
6.	Minimum temperature of the coldest month (°C)
7.	Temperature annual range (°C)
8.	Mean temperature of wettest quarter (°C)
9.	Mean temperature of the driest quarter (°C)
10.	Mean temperature of the warmest quarter (°C)
11.	Mean temperature of the coldest quarter (°C)
12.	Annual precipitation (mm)
13.	Precipitation of the wettest month (mm)
14.	Precipitation of the driest month (mm)
15.	Precipitation seasonality (CV)
16.	Precipitation of the wettest quarter (mm)
17.	Precipitation of the driest quarter (mm)
18.	Precipitation of the warmest quarter (mm)
19.	Precipitation of the coldest quarter (mm)
20.	Annual mean radiation
21.	Highest periodic radiation
22.	Lowest periodic radiation
23.	Radiation seasonality (CV)
24.	Radiation of the wettest quarter
25.	Radiation of the driest quarter
26.	Radiation of the warmest quarter
27.	Radiation of the coldest quarter
28.	Annual mean moisture index
29.	Highest period moisture index
30.	Lowest period moisture index
31.	Moisture index seasonality (CV)
32.	Mean moisture index of the highest quarter moisture index
33.	Mean moisture index of the lowest quarter moisture index
34.	Mean moisture index of the warmest quarter moisture index
35.	Mean moisture index of the coldest quarter moisture index

A range of landscape attributes was relevant only to the 86 eucalypt remnants and included:

• The distance between a given remnant and the nearest 1,000 ha block of *Eucalyptus* forest;

- The area of native forest within a circle of 1,000 m radius centred on the midpoint of a given remnant;
- Area-to-perimeter ratio of each remnant.

Digital information on contours, roads and drainage lines was obtained for each site from computer-based land cover surfaces. These data enabled the calculation of measures such as stream order (*sensu* Strahler, 1957), upslope catchment area and a range of other terrain and catchment attributes. Further details on these and other similar types of variables, as well as the methods used to measure them, are summarised in Table 4.6.

The third broad category of variables included measures of the structure and plant species composition of the vegetation gathered at all 166 sites in the cross-sectional study. These were measured in 20 × 20 m plots at 100 m intervals along the flagged transect for each site. Vegetation structure and plant species composition data were gathered because other studies of vertebrates have indicated that patterns of animal distribution

Table 4.6. *Summary of landscape variables collected for eucalypt remnants surrounded by stands of Radiata Pine and the methods used to measure them.*

Variable	Description
Perimeter	The perimeter of each patch. Extracted from a GIS database derived for the plantation and associated remnants
Perimeter-to-area ratio	The ratio of the area of a remnant to its perimeter
Distance to native forest	The distance (measured in metres) from the edge of a patch to the nearest large consolidated area (>1,000 ha) of native *Eucalyptus* forest
Pine age	The age of the Radiata Pine forest surrounding a given patch and reflecting the number of years since the original native forest had been cleared and replaced by exotic conifer trees. Calculated from stand class maps developed for the plantation by State Forests of NSW
Surrounding vegetation	The type of vegetation surrounding a remnant was classified as 'heterogeneous' if there was marked variation (>20 years) in the ages of the adjacent Radiata Pine (such as a clearfall on one edge of a remnant and a clear stand on the other). The type of surrounding vegetation was 'uniform' if the entire area comprised stands of Radiata Pine of broadly similar age (i.e. <20 years difference in the age of Radiata Pine forest)
Connectivity	Remnant was isolated from other areas of native forest (= 'discrete') or linked to other areas of eucalypt forest (= 'non-discrete')

and abundance may be strongly influenced by these factors. In the case of remnants 1–3 ha in size, the number of vegetation plots measured depended on the size of the site and thus the overall length of the flagged transect (see above). In all other cases, a total of seven plots was completed. Measured variables are described in Table 4.7. To facilitate statistical analyses, data at the plot level were usually aggregated to give a mean value for the site. As in the case of climate variables, there was collinearity between some of the plot attributes. Given this, the multidimensional

Table 4.7. *Plot-level vegetation structure, plan species composition and other variables.*

Variable	Description
Dominant tree species	The dominant species of trees in a remnant (identified from buds and fruits)[a]
Stand basal area	Measured in m^2 per ha using a basal area wedge
Aspect and topography	The topographic position of a site, in one of six categories: flat, gully, north-facing slope, east-facing slope, south-facing slope and west-facing slope
Disturbance	Evidence of disturbance was recorded and classified as mining, grazing, fire, logging and none
Dieback index	Evidence of dieback among dominant trees was recorded (e.g. crown and/or lateral branch death)
Hollow trees	The abundance of trees with bayonet and branch hollows was recorded[b]
Slope angle	The inclination of a plot measured using a clinometer
Bark Index	The quantity of strips of decorticating bark peeling from the trunk and lateral branches of trees[c]
Foliage depth	The depth of foliage from the top to the bottom of the largest tree crown in the plot (measured in m)
Rock index	A rock cover index was recorded for each plot as one of six classes[d]
Bracken index	The percentage cover of Austral Bracken (*Pterideum esculentum*) on the forest floor was recorded as one of six classes[d]
Ground cover	The percentage cover of the forest floor was assigned to one of six classes
Number of logs	The number of eucalypt and softwood logs in each diameter classes was recorded (10–20 cm, 20–30 cm, 30–40 m, 40–50 cm and >50 cm
Windrows	The number of, and distance to, windrowed piles of flogs remaining from the previous stand of *Eucalyptus* forest that was cleared and burnt to establish Radiata Pine trees was recorded. This measurement was pertinent only to Radiata Pine sites
Dominant cover	The percentage cover of dominant trees was recorded as one of six classes[d]

Table 4.7. (*cont.*)

Variable	Description
Sub-dominant cover	The percentage cover of sub-dominant plants was recorded as one of six classes[d]
Shrub cover	The percentage cover of shrubs was recorded as one of six classes[d]
Grass index	The percentage cover of grass was recorded as one of six classes[d]
Litter depth	Ranked in four categories from none to high
Blackberry index	An index describing the prevalence of introduced *Rubus fruticosus* scored from none to 100% in six categories each of 20% intervals
Plant matrix	A two-way height and diameter matrix was completed for each plot. Each stem in the plot was assigned by one of five height and seven diameter classes. The height categories were 1–2 cm, 5–10 cm, 10–20 cm, 20–30 cm, 30–40 cm, 40–50 cm and >50 cm
Pine invasion index	Index (0–5) of the extent of Radiata Pine wildlings established along the transect (not relevant in Radiata Pine sites)
Pruning status	The stage in the silvicultural cycle of plantation Radiata Pine was recorded. The amount of pruning slash on the forest floor (e.g. lateral branches remaining from thinned trees) was also recorded. These data were relevant to Radiata Pine sites only
Geology	Predominant lithology derived from geological surveys and mapping completed in the Tumut region
Stream index	A measure of the 'moistness' of a site calculated from information on the distance to a watercourse and stream order.[e] The index was calculated by dividing the distance to the closest stream from the start of a given transect by the order of the stream. This was repeated from the mid- and endpoint of each transect. The three values were then added to give a single overall number for the stream index
Eucalypt species present	Number of different species of eucalypts recorded (summed across all plots)

[a] See Costermans (1994); [b] *sensu* Jacobs (1955); [c] see Lindenmayer *et al.* (1990); [d] the six classes were none, 1–5%, 5–15%, 15–20%, 30–60% and >60%; [e] *sensu* Strahler (1957).

scaling procedures outlined above were employed to overcome these difficulties.

Final experimental design

In summary, 166 sites were selected for field surveys of biota – 86 sites in eucalypt remnants surrounded by extensive stands of Radiata Pine

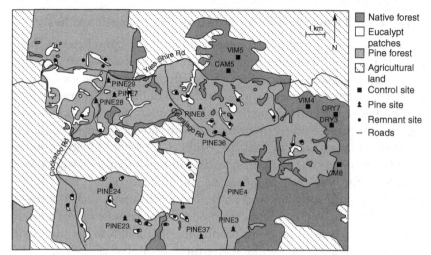

Figure 4.4. A subsection of the study region with sites in the three broad context classes (redrawn from Lindenmayer *et al.*, 2002a).

(Landscape Context Class 1), 40 sites dominated by Radiata Pine trees which formed the landscape matrix in the study region (Landscape Context Class 2), and 40 sites in large continuous areas of native eucalypt forest (Landscape Context Class 3). A subsection of the study area showing examples of the three broad types of sites is shown in Figure 4.4. In addition to the 166 selected sites, a further 41 micro-patches were surveyed for some taxa, such as arboreal marsupials and birds.

Reasons why the Tumut area was selected for study

Between 1992 and 1994 a large number of regions first seen in 1992 were examined and appraised as possible areas for detailed study. The Tumut region was selected in 1995 for a suite of reasons. These are set out briefly below.

The availability of 'control sites'

In human-modified landscapes, it is possible that a species may be absent from a set of remnants, not as a result of a landscape change effect but rather because it may never have occurred in such areas for other reasons, such as the unsuitability of climatic conditions. Thus, the use of 'control' areas could help to establish whether a given species would be expected to occur in a particular broad vegetation type. In the case of regions

---

Box 4.2. Other reasons for 'control' sites: genetic studies

The existence of large areas of native forest at the boundaries of the plantation at Tumut was important for several kinds of studies at Tumut such as those associated with the effects of landscape modification on patterns of genetic variability. Some of the most robust kinds of studies of landscape genetics involve contrasts in various kinds of gene metrics in modified environments with those in contiguous areas somewhat unmodified by humans (Banks *et al.*, 2005a, 2005b, 2005c; Taylor *et al.*, 2007). For example, as will be shown in Chapter 9, limited gene flow was found in the populations of the Bush Rat inhabiting remnant patches of native forest. However, such patterns were **not** an outcome of landscape modification, because similar levels of restricted gene flow were found among Bush Rat populations sampled in areas of contiguous native forest (Peakall *et al.*, 2003).

---

that encompass some human-modified landscapes, 'control' areas could be places that support similar vegetation types to those in a network of remnants but which are located in an 'unfragmented' spatial setting. The absence of a species from remnant vegetation of a similar type may, in turn, provide evidence that factors other than environmental suitability have led to its demise. True control areas have often not been included in landscape-scale studies of landscape change and habitat fragmentation, although there are some notable exceptions (e.g. Margules, 1992; Davies *et al.*, 2000).

The potential to cross-match eucalypt remnant and control sites

The human modification of landscapes has been quite systematic within virtually all of the regions in south-eastern Australia examined in the preliminary reconnaissance phase of this project. For example, vegetation on flat terrain and on high-quality soils had often been cleared or highly disturbed. Such a phenomenon is well known throughout many parts of Australia (Pressey, 1995; Landsberg, 1999) and indeed around the world (e.g. Scott *et al.*, 2001; Groom *et al.*, 2005). As a consequence, soil and vegetation types, as well as climatic regimes, which characterise remnant vegetation often contrast markedly with those typical of large areas of relatively undisturbed native vegetation such as National Parks, nature reserves and State Forests (Pressey *et al.*, 2000). Indeed, potential control areas in the regions examined in the preliminary phase of this project were either rare or did not exist. However, in the Tumut region, a number of

State Forests, together with the northern end of Kosciuszko National Park, supported extensive areas of native forest similar to the eucalypt remnants embedded within the Radiata Pine plantation. This made it possible to select sites which could be matched with the remnants on the basis of potentially important factors such as vegetation type and climatic regime. Moreover, the availability of large datasets for the Tumut region enabled us to examine patterns of spatial variation in environmental conditions such as aspect, slope, vegetation cover and bioclimatic regimes. These datasets greatly assisted the identification of areas containing suitable areas for use as 'control' sites.

Known history of landscape modification

Many studies, such as those by Suckling (1982), Loyn (1987) and Stouffer *et al.* (2006), have demonstrated that time since isolation is an important significant explanatory variable in explaining the suite of species lost from patches of remnant native vegetation within landscapes modified by humans. In the case of the Tumut region, remnant areas of forest have been isolated within the Radiata Pine plantation for periods of 10–60 years. Thus, a considerable period of time, and variation in isolation times, characterised our study region, which enabled us to examine the potential impacts of this variable on species persistence.

The disturbance history of many fragmented landscapes is often poorly or imprecisely known, making it a difficult factor to incorporate fully into subsequent analyses. However, in the case of the Tumut region, data were available on the precise time (to the month) that any new compartment of Radiata Pine forest was established. Thus, the period of time at which eucalypt remnants were surrounded by areas of Radiata Pine was accurately calculated.

The relative uniformity of landscape (pine) 'matrix' sites

The selection of the Tumut region was influenced by the fact it was relatively easy to survey fauna in the Radiata Pine plantation within which areas of remnant native vegetation were embedded. These areas are relatively homogeneous, and data about the current status of these areas (e.g. age, thinned vs. unthinned, etc.) were readily available from the government agency responsible for the management of the plantation. Inclusion of sites located in stands of Radiata Pine was critical. Many studies of the biota in human-modified landscapes have focused almost exclusively on areas of remnant native vegetation and ignored the value of the

---

Box 4.3.   Eight important design features: a summary of Tumut's advantages

Eight important design features characterise the main study at Tumut and these, in turn, provide a strong inferential basis for the interpretation of field results. These design features are:

(1) Sites in large areas of continuous eucalypt forest provided 'control' areas. These areas cover tens of thousands of hectares – large enough to support viable populations of the array of species which should occur in the study region.

(2) The 40 sites in Radiata Pine stands ensured that the value for biota of the landscape matrix surrounding eucalypt remnants could be properly accounted for (see also Simberloff *et al.*, 1992; Beier and Noss, 1998).

(3) Stratification for the selection of the eucalypt remnants ensured that the full environmental space of the study region was represented.

(4) Random selection within each stratum for the selection of the 86 eucalypt remnants minimised the chance of bias and allowed averaging over random factors.

(5) Climate, forest type and geology data ensured that the range of environmental and other conditions was matched across the three landscape context classes.

(6) Experimental units were of fixed size (a 600 m-long transect).

(7) Sampling intensity was independent of remnant size, which was appropriate for a comparative investigation (i.e. animal occurrence in patches was not confounded with sampling intensity).

(8) Field survey protocols for various groups (e.g. birds and arboreal marsupials) were established following a pilot study (Lindenmayer *et al.*, 1997a; Cunningham *et al.*, 1999; Chapter 5).

---

surrounding landscape for biodiversity (reviewed by Lindenmayer and Fischer, 2006). Until recently, relatively few studies have examined the fauna of both the remnant areas and the surrounding landscape (Ricketts, 2001; Lindenmayer and Franklin, 2002). However, the surrounding landscape may support populations of some species also found in the remnants, and they may make an important contribution to population persistence (Laurance, 1991; Andrén, 1994). Moreover, sampling of the surrounding landscape may reveal that some areas considered to be 'remnants' are in fact part of a habitat continuum and are not acting as a suite of habitat patches at all; in these cases the entire landscape is suitable for use by a given species (Lindenmayer and Fischer, 2006).

Previous understanding of faunal populations

The final reason why the Tumut region was an attractive place to work was that there had been a number of past investigations of the fauna in the area (e.g. Smith, 1969; Tyndale-Biscoe and Smith, 1969a, 1969b; Marples, 1973; Gall, 1982; Caughley and Gall, 1985; Rishworth *et al.*, 1995a, 1995b). These studies provided valuable background information on the species which could occur in the region.

Limitations of the cross-sectional study of landscape context effects

The preceding sections indicate that considerable thought and effort was dedicated to the design and implementation of the main cross-sectional study at Tumut. Despite this, the work was not without its problems. For example, as with the vast majority of studies of landscape change and habitat fragmentation, data were unavailable on the fauna of the different patches of remnant native vegetation at Tumut before plantation development began to occur (in 1932). Hence, it was not possible to document changes that may have occurred over time. This limitation was taken into account in the interpretation of findings from field surveys at Tumut. A subsequent major longitudinal study in the neighbouring Nanangroe region has attempted to overcome this problem (Lindenmayer *et al.*, 2008a; Chapter 10).

In some cases, the design of the major cross-sectional study at Tumut was unsuited to addressing other important questions of particular ecological and/or management interest or for attempting to quantify the responses of particular biotic groups to landscape context effects. In these cases, a new study had to be designed and implemented that often focused on particular subsets of sites. Some of these additional studies, and the key findings from them, are described in Chapter 10.

Target groups selected for study

Many studies have shown that different biotic groups can vary in their responses to landscape change and habitat fragmentation (Robinson *et al.*, 1992; Debinski and Holt, 2000; Lindenmayer and Fischer, 2006).

Conclusions drawn from one group may be poor indicators of the responses of other groups. Given this, a conscious decision was made to work on a range of biotic groups. While birds and mammals were the main focus of field research, other groups were investigated, including reptiles (Fischer et al., 2005), frogs (Parris and Lindenmayer, 2004), vascular plants (Smith, 2006), invasive plants (Lindenmayer and McCarthy, 2001), bryophytes (Pharo et al., 2004) and invertebrates (Smith, 2006).

Different field methods were employed to survey different groups, and often subsets of sites were used, rather than all 166 sites in the main cross-sectional study. For example, it was logistically impossible to survey all sites for invertebrates and bryophytes and only a small subset could be targeted. In the case of frogs, additional sites that encompassed wetlands had to be added to better capture the environmental space relevant to this group (Parris and Lindenmayer, 2004). In several cases, different survey protocols were employed for the same group (e.g. for birds; Lindenmayer et al., 1997b, 2004a; Cunningham et al., 1999, 2004a), and this informed the interpretation of results.

Although a wide range of groups were studied at Tumut, some have still not been investigated. Indeed, few parts of the earth have received a comprehensive study of all their biotic elements (Gaston and Spicer, 2004). For example, there are clearly many more invertebrate groups than the three broad ones examined by Smith (2006). Funds to support taxonomic work are a limiting factor for these kinds of invertebrate research, although future postgraduate research may help overcome this problem (see Chapter 12).

Bats are another group notably absent from the groups studied at Tumut. There is a good reason for this. Bats move large distances (often exceeding tens of kilometres) between roost sites and foraging areas (Lumsden et al., 1994; Lumsden and Bennett, 2005). This makes it hard to interpret the meaning of captures or other kinds of detections of these animals. Are animals that are trapped near their roost sites, on their way to foraging places or actively foraging in an area? Moreover, the scale of bat movements could exceed the size of much of the plantation at Tumut. Hence, the scale of movements may well be mismatched to the scale of the experimental design being examined.

Summary

The foundation of the Tumut Fragmentation Study was a major cross-sectional study of vertebrates in three landscape context classes – patches of remnant native forest surrounded by pine stands, matched sites in areas of contiguous native forest and areas dominated by stands of Radiata Pine.

A total of 166 sites was selected in these three landscape context classes using an iterative process leading to a stratified randomised and replicated site selection procedure. Repeated surveys of a range of different groups in these sites (or a subset of them) were completed. The results of these surveys are the primary focus of the following chapter.

5 · *The core findings: the effects of landscape context on animals and plants*

This chapter summarises the results of studies of landscape context effects for nine groups of animals and plants. The description of findings for each group is preceded by an outline of how each group was surveyed – the counting method and the number of sites surveyed was different in each case. The final part of this chapter contrasts the results for different groups and reflects on the main lessons of the work.

Survey methods

All empirical studies of landscape change and habitat fragmentation are underpinned by the quality of the field data gathered. Data quality is, in turn, influenced by the field methods employed. Failure to consider the efficacy of such methods may mean that what are perceived to be the effects of landscape change and habitat fragmentation are, in fact, artefacts of the field survey techniques.

A sub-theme at Tumut was therefore to test the accuracy and effectiveness of methods for sampling the major groups of vertebrates targeted in the study, particularly terrestrial mammals, birds and arboreal marsupials. These data were used to assist in the interpretation of field data or to modify field counting protocols to ensure that the best quality data were gathered. In addition, different methods and/or different subsets of the 166 sites were surveyed to estimate presence, abundance or cover. For example, it was logistically impossible to survey all 166 sites for invertebrates and bryophytes, and hence a subset of sites was selected. Similarly for frogs, there was little point in surveying terrestrial areas for this group and work targeted those sites that contained a water body (Parris and Lindenmayer, 2004). Given these considerations, a short discussion on field survey methods and site selection precedes the overview of landscape context effects for all groups.

Arboreal marsupials

Methods

Field data on arboreal marsupials were gathered by spotlighting. A field calibration study was completed to test the effectiveness of the method to detect the Greater Glider – the species of arboreal marsupial thought to be one of the most readily detected by spotlighting. The location of animals revealed by spotlighting was compared with the precise locations of a known population of radio-collared individuals (determined simultaneously by another observer using radio-telemetry equipment and working independently from the spotlight observer). The calibration study showed that spotlighting can seriously underestimate the abundance of the Greater Glider – sometimes by more than 50%. However, the extent of such bias was similar in a given forest type and the spotlighting method was considered to be valid for comparisons between 'treatments' such as patches dominated by similar types of native eucalypt forest (Lindenmayer et al., 2001a).

Arboreal marsupials were recorded along the 0–600m permanent transect set up at each of the 166 sites. Spotlighting was confined to still, clear nights to further minimise potential confounding of weather on the results. In addition, field work was terminated before 1.00 a.m. to limit the effects of observer fatigue, which could result in animals being missed. Potential confounding of landscape context type effects was limited by including at least one representative of each of the three broad classes of sites on any given night of the spotlighting survey. Finally, two observers completed all spotlighting surveys. Calibration work (see Chapter 4) showed that there were no significant observer differences in the detectability of different species of arboreal marsupials or the total number of animals per se (Lindenmayer et al., 1997a).

Results

Substantial landscape context effects were identified for arboreal marsupials (Figure 5.1). This finding was partly a consequence of the almost complete absence of most species from Radiata Pine stands. This was most probably due to the shortage of trees with hollows in pine stands – a key resource without which almost all species of arboreal marsupials cannot survive (Gibbons and Lindenmayer, 2002). In addition, an absence of flowers, fruits and invertebrate prey would have limited the suitability of foraging substrates for arboreal marsupials in stands of Radiata Pine.

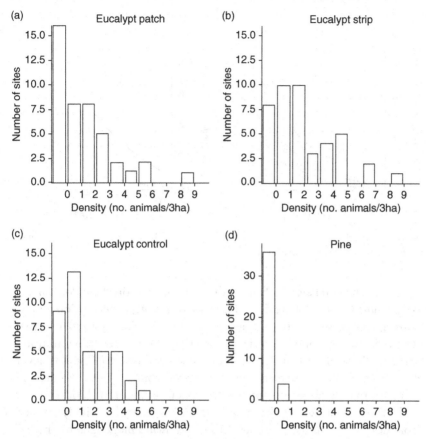

Figure 5.1. The density of arboreal marsupials in sites in different landscape context classes (redrawn from Lindenmayer *et al.*, 1999a).

The arboreal marsupial assemblage typical of remnant patches of eucalypt forest surrounded by Radiata Pine forest was different from that in a site embedded within extensive continuous eucalypt forest. Two species – the Yellow-bellied Glider (*Petaurus australis*) and the Squirrel Glider (*Petaurus norfolcensis*) – were absent from all the remnants, including the very large ones (greater than 80–120ha). Possible reasons for the loss of these species were that the Squirrel Glider was rare at the time of plantation establishment and the Yellow-bellied Glider has a very large home range (up to 60ha), which was bigger than the size of most patches (Lindenmayer *et al.*, 1999a).

Of the more common species of arboreal marsupials, the strongest landscape context and patch size effects were observed for the two species

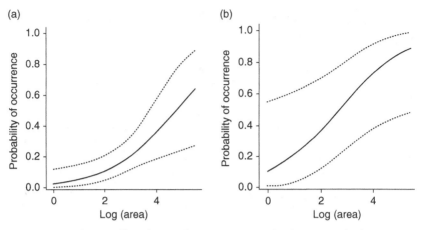

Figure 5.2. Patch size effects for: (a) the Common Brushtail Possum; (b) the Greater Glider (redrawn from Lindenmayer *et al.*, 1999a).

totally absent from Radiata Pine stands, the Common Brushtail Possum and the Greater Glider (Figure 5.2) (Lindenmayer *et al.*, 1999a). Conversely, landscape context effects were less pronounced for the two species that persisted at small population sizes in Radiata Pine stands (the Common Ringtail Possum and the Mountain Brushtail Possum) (Lindenmayer *et al.*, 1999a). Both these species were more abundant in the eucalypt remnants than in large areas of contiguous eucalypt forest (Lindenmayer *et al.*, 1999a) – an effect that would not be forecast by widely applied ecological paradigms (see Chapter 2) such as 'island biogeography theory' (MacArthur and Wilson, 1963, 1967) or 'metapopulation theory' (Hanski, 1999).

The pattern of occurrence of arboreal marsupials in eucalypt forest (i.e. the eucalypt remnants and areas of contiguous native forest) was significantly influenced not only by landscape context but by dominant forest type, vegetation structure and the steepness of the terrain. In addition, for the eucalypt remnants, patch size was important. Thus, vegetation quality as well as patch size strongly influenced the distribution of arboreal marsupials in the eucalypt remnants. In general, animals were more likely to occupy larger patches dominated by Ribbon Gum and Narrow-leaved Peppermint stands and on flatter terrain (Lindenmayer *et al.*, 1999a). However, each taxon exhibited a different (species-specific) response and, for some (e.g. Common Ringtail Possum and Mountain Brushtail Possum), factors such as remnant size were **not** important.

Additional analyses using indices of landscape cover (see also Chapter 7) revealed that populations of arboreal marsupials in remnants were larger in those that were isolated than in remnants with other eucalypt patches in the surrounding Radiata Pine stands. This may have been because Radiata Pine-dominated stands were hostile to dispersal and encouraged animals to remain within the patch in which they were born. This, in turn, would increase population densities in isolated remnants above those in less isolated remnants where animals may have been less reluctant to disperse (Lindenmayer et al., 2002b; Pope, 2003).

An interesting outcome of the work on landscape context effects was the co-occurrence of the closely related Greater Glider and the Common Ringtail Possum in 11 of the 40 sites within contiguous areas of native forest. However, co-occurrence was recorded in only two sites within eucalypt remnants. The lack of co-occurrence was most pronounced for eucalypt remnants dominated by Ribbon Gum and Narrow-leaved Peppermint, suggesting a habitat quality influence on combined patch occupancy for these two species.

Small terrestrial mammals

Methods

Two field sampling methods were employed in surveys of terrestrial mammals – aluminium 'Elliott' box trapping and hair-tubing (Lindenmayer et al., 1999c, 1999d). Bait placed inside an Elliott trap entices an animal to enter the device, and captured animals can be marked and released, providing standard abundance data in a mark–recapture form. Hair-tubing is a technique for detecting animals from the analysis of fur collected in a small portable bait station (Figure 5.3). Data are detection information only, as without DNA analysis (see Taylor et al., 1998), it is not possible to determine whether a single animal has visited a hair-tube multiple times or whether several different individuals have entered the device once.

A comparison of trapping and hair-tubing methods was completed as part of field surveys by establishing many hundreds of field plots with both a single Elliott trap as well as several types of hair-tubes (e.g. devices varying in entrance diameter). Patterns of concurrent detection of a given species by trapping and different types of hair-tubes were then examined. The method which produced the highest number of detections varied between species. For small native mammals such as the Bush

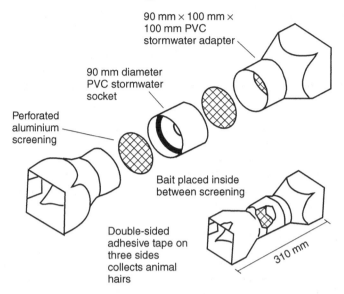

90 mm × 100 mm × 100 mm PVC stormwater adapter

90 mm diameter PVC stormwater socket

Perforated aluminium screening

Bait placed inside between screening

Double-sided adhesive tape on three sides collects animal hairs

310 mm

Figure 5.3. A standard kind of hair-tube used in detecting mammals at Tumut.

Rat and *Antechinus* spp., trapping was significantly more effective than hair-tubing for detecting animals (Lindenmayer *et al.*, 1999c, 1999d). There were many plots where these species were trapped but never recorded in any of the four types of hair-tubes trialled. This highlighted the importance of accounting for survey methodology in interpreting landscape context effects.

Two major surveys of terrestrial mammals were completed at Tumut. Hair-tubing surveys were conducted at all 166 sites selected in the study area. A survey involving a combination of different types of hair-tubing and Elliott trapping was completed in a subset of 58 sites. Data from these surveys were used to investigate the response of mammals to landscape context and other attributes.

Results

Small terrestrial mammals were recorded significantly less frequently in Radiata Pine sites than in sites in large continuous areas of eucalypt forest or eucalypt remnants surrounded by the softwood plantation. Piles of windrowed eucalypt logs remaining after clearing of the original native forest were the only locations within the Radiata Pine stands where these species were recorded (Lindenmayer *et al.*, 1999c).

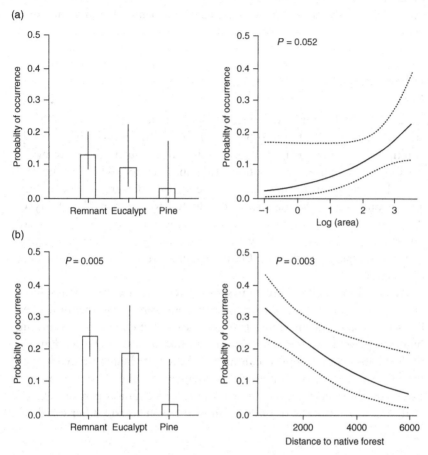

Figure 5.4. (a) Landscape context and patch area effects for the Bush Rat; (b) Landscape context and patch isolation effects for the Agile Antechinus (redrawn from Lindenmayer *et al.*, 1999c).

Strong positive relationships were also observed between the remnant area and the probability of patch occupancy by the Agile Antechinus and Bush Rat; small patches were generally those most likely to be empty. In the case of the Agile Antechinus, isolated eucalypt patches were also least likely to be occupied (Figure 5.4) (Lindenmayer *et al.*, 1999c).

As in the case of arboreal marsupials, the occurrence of the Bush Rat and the Agile Antechinus was associated with some forest attributes. These included stand basal area, the steepness of the terrain and the dominant (eucalypt) tree species (Lindenmayer *et al.*, 1999c). Together with landscape context effects, these attributes highlighted the influence of factors at several spatial scales on the occurrence of the Bush Rat and the Agile Antechinus.

Birds

Methods

Bird surveys at Tumut involved repeated point interval counts (*sensu* Pyke and Recher, 1983) at the 166 sites in the main cross-sectional study. Multiple volunteer observers were required to complete these surveys to ensure that sampling was confined to a short period when calling and other behavioural patterns (which influence bird detectability) were relatively uniform. Field trials were conducted to compare the ability of different observers to count birds in different forest types (eucalypts versus pines) using different counting protocols (point–interval counts, zigzag walks and straight transects) (Lindenmayer *et al.*, 1997b; Cunningham *et al.*, 1999).

Results showed that there were significant differences between observers in their ability to detect particular groups of birds (e.g. medium-sized birds that forage in the understorey and call frequently) – even among highly experienced observers (Cunningham *et al.*, 1999). To account for this problem, counting protocols involved each site being surveyed twice on different days by two different experienced observers. In addition, the extent of observer differences was taken into account as part of detailed analyses of the response of birds to landscape context and other variables.

Results

The forest mosaic at Tumut supported more than 90 species of birds (Lindenmayer *et al.*, 2007c; see Appendix 2) and complex landscape context and other effects were identified. Bird species richness was lowest in stands of Radiata Pine and highest in large areas of native eucalypt forest (Figure 5.5). Patches of remnant native forest were characterised by intermediate levels of species richness (Lindenmayer *et al.*, 2002a).

More complex landscape context effects were found at the individual species level. Some taxa were ubiquitous, such as the Grey Shrike Thrush (*Colluricuncla harmonica*). Another group of taxa were recorded most often in the remnants; e.g. Little Raven (*Corvis mellori*), Superb Fairy Wren (*Malurus cyaneus*) and Shining Bronze Cuckoo (*Chyrysococcyx lucidis*). Of these, some – e.g. the Australian Magpie (*Gymnorhina tibicen*) – were more likely to be recorded in small eucalypt remnants; whereas the Eastern

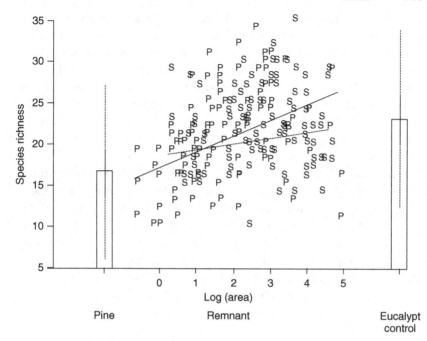

Figure 5.5. Gradient in landscape context effects for bird species richness (redrawn from Lindenmayer *et al.*, 2002a).

Yellow Robin (*Eopsaltria australis*) and Superb Lyrebird (*Menura superba*) were more likely to be found in the intermediate-sized eucalypt remnants; and the Red Wattlebird (*Anthochaera carunculata*), Sacred Kingfisher (*Todiramphus sanctus*), Leaden Flycatcher (*Myiagra rubecula*) and White-naped Honeyeater (*Melithreptus lunatus*) in larger remnants. Other birds which typically favoured large continuous areas of native forest were the Cicada Bird (*Coracina tenuirostris*), Gang-gang Cockatoo (*Callocephalon fimbriatum*) and Olive-backed Oriole (*Oriolus sagittatus*). A number of native species were significantly more likely to occur in Radiata Pine stands, e.g. the Rufous Whister (*Pachycephala rufiventris*) and the Brown Thornbill (*Ancathiza pusilla*).

Analyses of the composition of the bird assemblage at each site were based on a combined measure of species presence and abundance (termed a 'bird frequency profile'). There was not a fixed community of birds in the three broad landscape context classes, or a climax bird community, but rather a complex reassembling of the composition of the bird assemblages in relation to landscape context, remnant area and conditions in the landscape surrounding a given site.

There was strong empirical evidence for a gradient in the bird frequency profiles between Radiata Pine stands and large continuous areas of native eucalypt forest. Changes in the bird frequency profiles along this continuum encompassed changes both in the identity of the taxa in the assemblage and the relative abundance of each species between the two broad Radiata Pine and continuous eucalypt landscape context classes. The remnants connected the bird frequency profiles of these two forest types and the nature of the gradient depended strongly on remnant size.

As remnant size increased, the bird assemblage was increasingly similar to that characteristic of large continuous areas of native eucalypt forest and less like that of the surrounding Radiata Pine (Figure 5.6). However, the change in the bird frequency profile for the various types of sites was strongly influenced by the landscape conditions that surrounded the field sites. For example, the occurrence of many bird species in the Radiata Pine stands was significantly related to the amount of eucalypt forest surrounding these pine sites (Lindenmayer *et al.*, 2002a).

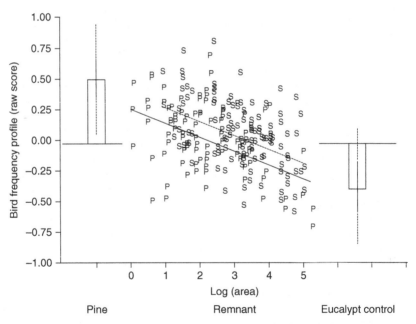

Figure 5.6. Gradient in landscape context effects for the bird frequency profile or a combined measure of overall species presence and abundance (see text). Redrawn from Lindenmayer *et al.* (2002a).

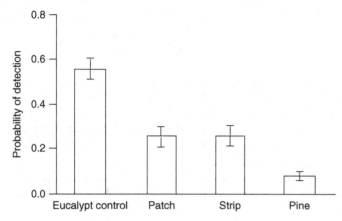

Figure 5.7. Radiata Pine–Remnant–Control eucalypt gradient for the presence and abundance of the Red Wattlebird (redrawn from Lindenmayer, 2000).

Pine sites where the surrounding landscape contained some eucalypt remnants had a different bird frequency profile from pine sites where the surrounding landscape was pure Radiata Pine. Radiata Pine sites with adjacent eucalypt patches had a bird frequency profile similar to that of small and intermediate-sized eucalypt remnants (Lindenmayer *et al.*, 2002a).

Many individual bird taxa at Tumut were distributed across the different landscape components at varying levels of abundance and these species were **not** confined to only one landscape context type. The Red Wattlebird provides a good example (Figure 5.7); the species was significantly more abundant in large continuous forest than in Radiata Pine stands and had intermediate levels of abundance in the eucalypt remnants.

An association was found between the life history attributes of species and their distribution across the different landscape context classes. Birds more likely to be detected in eucalypt forest (both contiguous eucalypt forest and eucalypt patches) belonged to particular foraging guilds: hawkers, bush carnivores, bark probers, wood searchers, foliage searchers, nectarivores and granivores. Those species which had a greater maximum nesting height were significantly more likely to occur in eucalypt forest than in stands of Radiata Pine. Significant relationships also were identified between the shape of the eucalypt remnants and life history attributes. Birds more often detected in elliptical-shaped eucalypt remnants were smaller, produced smaller clutches and were more likely to be migratory; in addition, these birds were more likely to have cup nests or burrows (Figure 5.8).

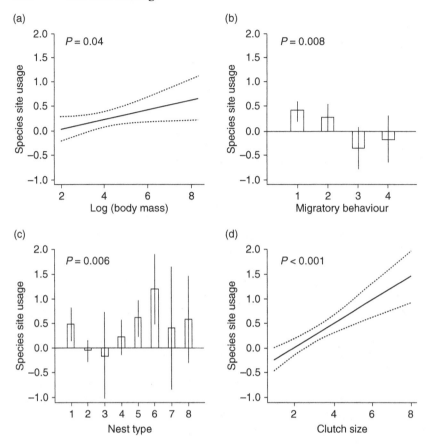

Figure 5.8. Relationships between life history attributes of birds and patches and strip-shaped eucalypt remnants. Patch-shaped remnants correspond to site usage scores of less than zero (redrawn from Lindenmayer *et al.*, 2002a).

As in the case of arboreal marsupials, analyses using landscape metrics for patterns of landscape cover showed that some taxa responded positively to the total area of remnant vegetation when it was dispersed among many eucalypt patches; e.g. Rufous Whistler and Crimson Rosella (*Platycercus elegans*). In contrast, other species were significantly more likely to occur if eucalypt patches were consolidated as a single (or small number) of large patches; e.g. Red Wattlebird and Golden Whistler (*Pachycephala pectoralis*) (Lindenmayer *et al.*, 2002a). The effects are discussed in more detail in Chapter 7.

Extensive analyses of patterns of spatial dependence in bird distribution (e.g. Koenig, 1998) were completed. These analyses showed that the distribution of many species did not conform to that expected if classic metapopulation dynamics were occurring (*sensu* Hanski, 1999), with particular taxa restricted only to a 'mainland' (here, the extensive native forest areas) or to certain patches or types of patches. Rather, Radiata Pine stands surrounding the remnants provided suitable or partially suitable habitat for many species.

Understorey conditions were statistically significant predictors of the presence and abundance of many species of birds. For example, occurrences of native understorey plants such as Dogwood (*Cassinia aculeata*) and Bracken Fern (*Pterideum esculentum*) were important for providing cover for species like the Brown Thornhill (*Ancathiza pusilla*) and the White-browed Scrub Wren (*Sericornis frontalis*). The presence of these structural features often meant that overall differences in species richness were less than expected between Radiata Pine stands (mean = 16 species), remnants (mean = 20 species) and large continuous areas of native forest (mean = 23 species) (Lindenmayer *et al.*, 2002a).

In summary, studies of birds at Tumut produced a range of interesting and novel results. First, small and intermediate-sized eucalypt remnants (between 1 and 10 ha) supported a wide range of species – and clearly have considerable conservation value.

Second, there were strong spatial interrelationships between Radiata Pine stands and remnant patches of eucalypt forest. Many taxa occurred in the Radiata Pine stands because of their adjacency to eucalypt forest.

Third, some of the responses to landscape condition and habitat fragmentation observed in the study (e.g. the preferential use of small and intermediate-sized patches by some native taxa) were quite different from those seen in other investigations of modified forest landscapes.

Reptiles

Methods

Fischer *et al.*, (2005) conducted extensive field surveys of reptiles within 30 sites at Tumut. Six of these were located in pine stands, three in recently clear-felled areas and three in 20-year-old stands. The remaining 24 sites were in eucalypt remnants and large areas of contiguous forest. The experimental design encompassed three forest types and four size classes (1–3 ha, 3–10 ha, >10 ha and controls (in contiguous forest)). Each forest type × size class combination was replicated twice (Fischer *et al.*, 2005).

At each of the 30 survey sites, four 20×10 m plots were established, giving 120 plots for the entire study. Two of the plots on a site were established in canopy gaps and two under tree canopies. This ensured that a range of thermal conditions that are important for reptiles was surveyed at each site. Each plot comprised a 20 m long × 30 cm high plastic drift fence with three 9.6 l pitfall buckets – one located at each end of the fence and one in the middle.

All sites were surveyed for reptiles on a repeated basis between November 2002 and February 2003. In addition to reptile data, Fischer *et al.* (2005) gathered site and landscape covariates as well as extensive life history information on the food and shelter requirements of each species of reptile captured in the study.

Results

Fischer *et al.* (2005) captured 13 species of reptiles – all of them skinks. Data analysis then involved exploring responses at the plot level and the site level for species richness, assemblage composition and individual taxa.

Species richness at the plot level increased significantly with the amount of eucalypt cover within 1,000 m. Site-level species richness was highest at intermediate values for elevation (Figure 5.9). The composition of the reptile assemblage turned over considerably between sites as a function of elevation, eucalypt cover within 1,000 m and the abundance of potential prey items such as springtails (Collembola) (Fischer *et al.*, 2005). Lizard species richness was highest at sites where the landscape context was

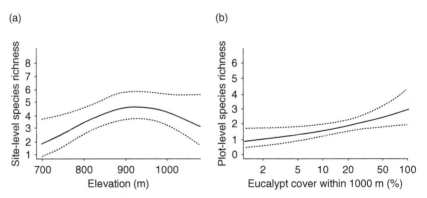

Figure 5.9. Reptile species richness responses at the site (a) and plot (b) levels. Diagrams show the predicted response (from a regression model; solid line) and the 95% confidence intervals (dashed line). Redrawn from Fischer *et al.* (2005).

dominated by native eucalypt forest. However, some species stands of reptiles also were found in stands of Radiata Pine, particularly those more than 20 years old (Fischer *et al.*, 2005).

Most of the work on reptiles involved building statistical models of the occurrence of individual species. Different suites of variables were significant in the statistical relationships identified for each taxon (Fischer *et al.*, 2005). Thus, each species showed individualistic responses to landscape and other attributes. These attributes could be typically assigned to broad categories of variables: climate, space, food and shelter, many of which operate at different spatial scales.

Space effects for reptiles typically related to the amount of native vegetation within 1,000m, reflecting landscape context effects rather than patch size effects. Importantly, responses to this variable varied in markedly different ways between different species. For example, Maccoy's Skink was more likely to be found at sites with a high value for the amount of native forest within 1,000m, whereas the opposite effect occurred for the Garden Skink (Fischer *et al.*, 2005).

Frogs

Methods

Parris and Lindenmayer (2004) used two field methods to survey frogs at Tumut. The first method involved the use of automatic call-box recorders to record frog calls for 20 minutes every hour between 6.00 p.m. and 9.00 p.m. for three consecutive days in February (summer) and October (spring) in 2000. This approach ensured that both diurnal and nocturnal calling frogs were detected, that between-night and between-season effects were taken into account and frogs that call in different seasons were included (Parris and Lindenmayer, 2004).

The 'call up' method (Lemckert, 1998) was the second technique used to survey frogs at Tumut. Some species of frogs call in response to a loud noise such as a person shouting. Call up by repeated shouting was conducted at each site in February 2001 and the numbers of each species of frog that responded were recorded (Parris and Lindenmayer, 2004).

The two field methods were employed in surveys at 60 of the 166 sites at Tumut which contained a water body – 29 in eucalypt remnants surrounded by pine stands, 15 in stands of Radiata Pine and 16 in contiguous areas of native forest.

Results

Parris and Lindenmayer (2004) found that frog species richness was lower in stream and wetland sites located within pine stands than at sites in contiguous areas of native forest or in patches of remnant native forest. Only two of the eight species of frogs recorded at Tumut occurred at water bodies where the immediate surrounding vegetation was dominated by Radiata Pine. The characteristics of water bodies had a significant effect on species richness. It was highest in wide, shallow swamps and marshes near the headwaters of streams, with herbs, grasses, reeds, shrubs, sedges and rushes dominating the emergent and fringing vegetation (Parris and Lindenmayer, 2004).

Frog species richness was **not** associated with the size of patches of remnant native vegetation surrounded by plantation pine stands. Isolation effects were also noticeably absent, suggesting that pine stands may not be a barrier to the dispersal of the eight species of frogs recorded at Tumut. However, Parris and Lindenmayer (2004) noted that there was low statistical power to detect trends for the effects of both patch size and patch isolation.

Invertebrates

Methods

Smith (2006) completed a major study of three groups of invertebrates at Tumut – Formicidae (ants), Diptera (flies) and Coleoptera (beetles). The work involved surveys of five kinds of sites:

(1) Stands of Radiata Pine;
(2) Areas of contiguous native forest;
(3) Small (1–10 ha) patches of eucalypt forest surrounded by Radiata Pine;
(4) Medium-sized (11–20 ha) patches of eucalypt forest surrounded by Radiata Pine;
(5) Large (>20 ha) patches of eucalypt forest surrounded by Radiata Pine.

Each site type was replicated five times, giving 25 sites in total. These were selected from the 166 sites that were established for work on vertebrates. All sites in eucalypt forests were dominated by Narrow-leaved Peppermint to negate the potential for differences in vegetation type to influence the results (Smith, 2006). The five sites in pine stands were all 25 years of age. Environmental variables such as soil type, geology and climate were similar for all 25 sites.

Invertebrates were captured in nine pitfall traps established within a 20×20 m grid on each of the 25 sites. The pitfall traps were 6.5 cm in diameter, 10 cm deep and contained ethanol and glycerol. All samples collected were then classified to morphospecies level (*sensu* Oliver and Beattie, 1996).

Results

The unpublished Ph.D. thesis by Smith (2006) contains a detailed presentation of the investigation of landscape context effects for invertebrates. Only a very brief overview of key findings is presented below.

In the case of ants, Smith (2006) identified marked landscape context effects. Ants were uncommon in stands of Radiata Pine and both species richness and the abundance of morphotaxa were lower than in other kinds of sites. Species richness was highest in sites located in large contiguous areas of native forest, and there was a trend for decreasing species richness with decreasing patch size (Table 5.1). In addition, the composition of ant assemblages in contiguous native forest was different from that of the eucalypt patches and the Radiata Pine sites (Smith, 2006).

In contrast to the results for ants, differences were not significant in the species richness, assemblage composition and overall abundance of flies between contiguous eucalypt forest and small and medium-sized patches (Smith, 2006). Species richness was lowest in stands of Radiata Pine and highest in the largest patches of eucalypt forest (Table 5.1).

No marked landscape context effects were identified for beetles. Smith (2006) found that species richness in stands of Radiata Pine was similar to that for contiguous native forest (Table 5.1). Beetle abundance was highest in the large eucalypt patches and the composition of the beetle assemblage in small eucalypt patches was different from the other kinds of sites.

In summary, different groups of invertebrates exhibited quite different responses to landscape context. Many of these differences appeared to be associated with the extent of use of Radiata Pine stands. Although stands of Radiata Pine were inferior environments for all groups, beetles and flies were relatively common and species-rich within them. Landscape context and patch size effects were limited for these groups. Ants were uncommon in pine stands and this group exhibited relatively strong landscape context effects. There was also a suggestion of patch size effects for this group.

Table 5.1. *Invertebrate species richness in different landscape context classes at Tumut (redrawn from Smith, 2006).*

Species group	Total N	Eucalypt forest			Large eucalypt patch			Medium eucalypt patch			Small eucalypt patch			Pine plantation		
		Total species	Mean species per site	P	Total species	Mean species per site	P	Total species	Mean species per site	P	Total species	Mean species per site	P	Total species	Mean species per site	P
Ants	68	38	16.2 2.27 SE	**0.048**	34	13.4 1.33 SE	NS	32	12 1.08 SE	NS	30	10.4 2.2 SE	NS	20	6.8 1.63 SE	**0.008**
Flies	106	53	21.2 2.2 SE	NS	68	26.0 1.67 SE	**0.016**	60	22.2 3.11 SE	NS	58	22.2 2.08 SE	NS	48	15.4 1.63 SE	**0.002**
Beetles	78	35	14.2 0.97 SE	NS	37	15.2 1.46 SE	NS	41	16.5 2.33 SE	NS	36	16.4 1.78 SE	NS	37	14.8 2.43 SE	0.62

Pooled species richness for each group of invertebrates (total N); pooled species richness for each group for each context class; mean species density per site for each species group; and the significance value (P) of the difference in mean species density for each class context against the mean for all other context classes. Significant values are given in **bold**.

Vascular plants

Methods

The experimental design for invertebrates (see above) was identical to that used by Smith (2006) in studies of landscape context effects on vascular plants. That is, plots within 25 sites comprising five replicates of each of five landscape context classes. Field surveys involved measuring the cover and abundance of all vascular plant species with a 20×20m plot located within each of the 25 sites

Landscape context effects for vascular plants

Smith (2006) identified 188 species of vascular plants in his study. Of these, 22 (12%) were exotic. Landscape context effects were identified at the species richness and community composition levels for native plant taxa. The trend, in order of decreasing species richness, was: contiguous native forest > large eucalypt patches > medium-sized eucalypt patches > small eucalypt patches > stands of Radiata Pine. However, the species richness of the Radiata Pine sites was unexpectedly high, with these areas supporting an average of 22% of the flora found in contiguous native forest (Smith, 2006). Broadly similar results were found for the composition of the vascular plant community. Radiata Pine sites had a composition that was distinctively different from both kinds of eucalypt sites. In general, the composition of large eucalypt patches was more similar to the contiguous native eucalypt forest than the medium- or small-sized eucalypt forests. Finally, native species of vascular plants that were found to be common in the pine sites were more common in the eucalypt patches than they were in sites within continuous areas of native forest.

Invasive vascular plants

Methods

Detailed information on vegetation structure and plant species composition was gathered at seven 20×20m plots along the 600m transect at all 166 sites in the landscape context study ($N = 1,162$ plots in total). Two key measurements from these extensive vegetation surveys were used to examine the problems of weed invasion. These measures were the presence/absence of Radiata Pine wildlings and the density of cover of Blackberry (*Rubus fruticosus*). Extensive statistical analyses were then completed of the factors influencing the occurrence of both invasive species (Lindenmayer and McCarthy, 2001).

Results

Lindenmayer and McCarthy (2001) found that pine wildlings were absent from plots within sites located in large areas of contiguous forest. However, some types of eucalypt remnants were particularly susceptible to invasion (Figure 5.10). Eucalypt remnants that had been surrounded by Radiata Pine stands for a prolonged period were more likely to contain pine wildlings. Pine wildlings also were significantly more likely to occur in dry forest and woodland types such as those dominated by Apple Box and Broad-leaved Peppermint.

Major landscape context effects were identified for the cover of Blackberry (Figure 5.10). Invasion levels varied from 59% for plots in stands of Radiata Pine, 41% for plots in eucalypt remnants to only 7% in contiguous areas of native eucalypt forest (Lindenmayer and McCarthy, 2001). In the case of the eucalypt remnants, occurrence increased significantly with proximity to the surrounding Radiata Pine stands. It was also more prevalent in wetter forest types, particularly gully systems dominated by Mountain Swamp Gum (Lindenmayer and McCarthy, 2001). Levels of Blackberry infestation were also greater in areas with

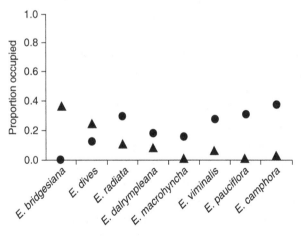

Figure 5.10. Predictions from the statistical relationship between the probability of occurrence of Blackberry (circles) and Radiata Pine wildlings (triangles) and eucalypt remnants dominated by different types of eucalypt forest (redrawn from Lindenmayer and McCarthy, 2001).

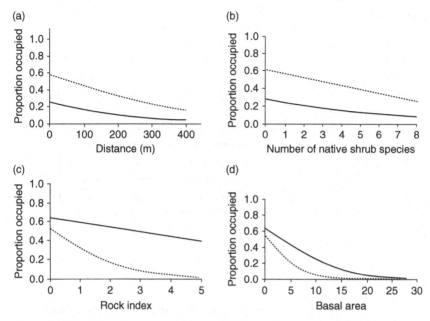

Figure 5.11. Relationships between the probability of occurrence of Blackberry and significant covariates of eucalypt remnants: (a) Distance to pine stands; (b) Number of species of native shrubs; (c) Rock index; (d) Basal area of eucalypts (redrawn from Lindenmayer and McCarthy, 2001).

fewer native shrubs, lower values for eucalypt basal area and limited rock cover (Figure 5.11).

The edges of eucalypt fragments closest to the pine plantation were where the heaviest infestations of Blackberry occurred; dense, 3 m–high thickets were common in these areas and most probably acted as large sources of propagules to promote the invasion of the adjacent eucalypt remnants. Recurrent high-intensity disturbance in pine plantations by logging machinery may also have enhanced conditions for the growth of Blackberry, and this may maintain source populations for weeds to invade nearby eucalypt remnants.

Analyses of landscape context effects for invasive species indicated that their prevalence may increase over time and that existing weed control options are probably limited at present. However, the adoption of some protocols governing the movement of machinery between existing and new plantations (which presently do not have severe weed infestations) could limit, or at least slow, their spread (Lindenmayer and McCarthy, 2001; Chapter 11).

Bryophytes

Methods

Pharo *et al.*, (2004) completed a study of landscape context effects on bryophytes. A subset of 32 of the 166 sites at Tumut was selected for detailed survey. These comprised 16 eucalypt remnants, 8 sites in large areas of contiguous forest and 8 sites in stands of Radiata Pine. The eucalypt remnants were divided between strip-shaped and elliptical-shaped sites.

Two kinds of surveys of bryophytes were conducted. First, a 10×10 m plot on each of the 32 sites was searched thoroughly and a record was kept of the substrate on which each species of bryophyte was detected (e.g. rock, log, tree trunk). Second, additional substrates (not found within the 10×10 m plot) were searched within an area of 100×50 m at each of the 32 sites (Pharo *et al.*, 2004).

Results

In total, 45 species of mosses and 13 species of liverworts were recorded at Tumut. Three key results were found. First, there were substantial landscape context effects driven, in part, by stands of Radiata Pine supporting 40% lower levels of species richness compared with sites in other landscape context classes (Pharo *et al.*, 2004). The eucalypt remnants were the most species-rich, with the sites in contiguous areas of native forest having intermediate values for species richness. No patch shape effects were identified for bryophyte species richness (Figure 5.12). Some highly unexpected patch size effects were identified, with the highest species richness in the intermediate size classes (3–10 and 11–20 ha) and the lowest in the smallest (<3 ha) and largest (>20 ha) patches of remnant eucalypt forest (Pharo *et al.*, 2004).

A second key result was that when the composition of the bryophyte flora was examined, a continuum of change in the species assemblage was found, grading from eucalypt remnants, continuous areas of native forest, through to stands of Radiata Pine (Pharo *et al.*, 2004). As in the case of analyses of species richness, the highest levels of contrast were between areas of eucalypt forest and stands of Radiata Pine. The bryophyte flora of strip-shaped remnants exhibited greater levels of similarity to that of contiguous areas of native forest than did elliptical-shaped remnants.

A third key result was that substrate type had a very strong effect on bryophyte occurrence. Different kinds of substrates supported different levels of species richness (Figure 5.13). They were also habitats for different species assemblages. Pharo *et al.*, (2004) found that bryophyte species

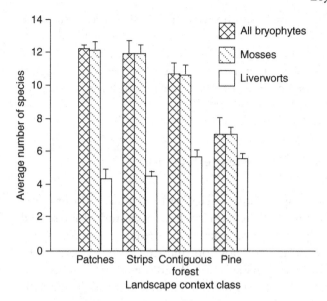

Figure 5.12. Landscape context and bryophyte species richness at Tumut (redrawn from Pharo *et al.*, 2004).

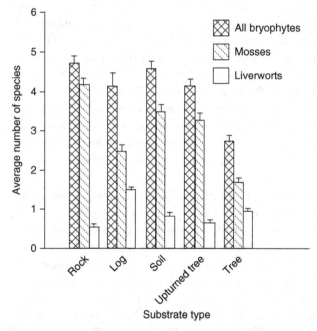

Figure 5.13. Bryophyte species richness associated with different kinds of substrates at Tumut (redrawn from Pharo *et al.*, 2004).

richness and assemblage composition was strongly influenced by substrate type. For example, many species characteristic of eucalypt forests were absent in pine stands. Of the depauperate flora found in pine stands, 90% of the species were associated with eucalypt logs that were the biological legacies remaining from clearing of the original native forest cover to establish the plantation (Pharo *et al.*, 2004).

Summary of landscape context effects for different groups

Quantification of landscape context effects has been a major thrust of the work at Tumut. A range of different groups have been studied and several levels of biological organisation have been examined for most groups (e.g. species richness, assemblage composition, individual species responses). This breadth of work has provided a much greater set of insights than would have been possible from a more limited investigation (Table 5.2).

Some of the important findings that have emerged to date are summarised below.

- The nature, magnitude and direction of landscape context effects vary markedly between broad groups as well as between assemblages and individual species within those groups. The effects (or the lack of them) could in some cases be associated with the occurrence in, and use of, stands of Radiata Pine. A good example was the contrast between effects for different invertebrate groups (Smith, 2006).
- Patch size effects for the eucalypt remnants were only sometimes observed and they were far from ubiquitous across groups (see also Debinksi and Holt, 2000). The strength of such effects often tended to be diminished for species and/or groups that were common and/or diverse in stands of Radiata Pine.
- Following on from the above two points, the human perspective of the plantation landscape at Tumut as 'islands of bush in a sea of pines' – or as patches, corridors and the pine matrix – was not a perspective shared by some species and/or broad groups. This has critical implications for choosing the particular conceptual model of landscape cover (see Chapters 2 and 7) which is the most appropriate to use. It is clear from the diversity of results that have been summarised in this chapter that no single model would be appropriate for general use across all the biotic groups that have been examined at Tumut (Fischer and Lindenmayer, 2007). The topic of testing different conceptual models for landscapes is revisited in Chapter 7.

Table 5.2. *Landscape context and patch size effects for different groups.*

Group	Landscape context effects	Comments	Patch size effects	Comments
Arboreal		marsupials	Yes	Almost all species absent from sites in pine stands
Yes	Not	apparent for all species		
Small		terrestrial mammals	Yes	Native species rare in pine stands
Yes	Largest patches most likely to be	occupied		
Birds	Yes	Many species absent from pine stands	Yes	Not apparent for all species
Reptiles	Yes	Reflected by measure of amount of native vegetation in surrounding 1,000 m	Rare	Climate, shelter and food covariates are important
Frogs	Yes	Most species absent from sites in pine stands	No	Attributes of water bodies critical
Ants	Yes	Ant fauna is depauperate in pine stands	Yes	Largest patches were most species-rich
Flies	Yes	Fly fauna is depauperate in pine stands	Partial	Species richness highest in large patches
Beetles	No	Few differences between pine stands and other context classes	Partial	Beetle abundance (but not species richness) highest in large patches
Vascular plants	Yes	~80% of vascular native flora missing from pine stands	Yes	Largest patches were most species-rich
Invasive plants	Yes	Invasive plants rare in contiguous native forest	No	Invasion levels of patches linked to age of surrounding pine stands
Bryophytes	Yes	Pine stands have limited bryophyte flora	Yes	Highest in intermediate patch sizes, substrate suitability critical

- Many species and/or groups responded to attributes at multiple spatial scales. That is, while landscape context effects may have been important, other effects such as the dominant vegetation type within eucalypt remnants or the availability of particular substrates (e.g. for reptiles and bryophytes) were also critical predictors of occurrence (Fischer *et al.*, 2005). Indeed, in the case of bryophytes, substrate type was of greater importance in accounting for occurrence than landscape context (Pharo *et al.*, 2004). Thus, a good understanding of the biology of a species is fundamental to predicting its response to landscape modification.

The focus of work on landscape context effects at Tumut has overwhelmingly been on pattern. That is, the pattern of animal or plant occurrence within sites in different landscape context classes. This is a natural outcome of a set of cross-sectional studies of this nature. However, many of the following chapters outline the results of studies that were focused more on attempting to better understand the mechanisms or processes giving rise to the patterns that have been described in this chapter.

6 · *Patch use: how animals use patches of remnant eucalypt forest surrounded by pine*

Landscape change can have direct negative impacts on species such as through the interconnected processes of habitat loss, habitat degradation and habitat subdivision (Lindenmayer and Fischer, 2006). Landscape change and habitat fragmentation can also lead to changes in the behaviour and biology of particular species (reviewed by Banks *et al.*, 2007). This, in turn, can alter patterns of patch use and, in some cases, result in population decline or even extinction (Simberloff, 1988; Banks *et al.*, 2007). As highlighted in the previous chapter, much of the work at Tumut has quantified patterns of site occupancy in different landscape contexts. However, other studies have either directly or indirectly explored the way that animals actually use patches of remnant native forest that were surrounded by stands of exotic Radiata Pine. This chapter summarises some of that work. It includes brief descriptions of:

- Studies of home range use in eucalypt patches by arboreal marsupials;
- Bird movements within patches as well as the surrounding pine stands;
- Bird calling behaviour within eucalypt patches surrounding Radiata Pine stands;
- Within-patch breeding biology of the small carnivorous marsupial, the Agile Antechinus.

Movement and other changes in patches of different sizes

Patch use by the Greater Glider

The size and spatial configuration of habitat patches in modified landscapes can have profound effects on their use by the individuals of a given species. For example, the home ranges and movement patterns of animals may be altered as a result of habitat subdivision (Arnold *et al.*, 1993; Brooker and Brooker, 2002). To determine whether such effects

Table 6.1. *Density and sex ratio of Greater Glider (*Petauroides volans*) in five patches of eucalypt forest (modified from Pope, 2003).*

Patch	Sex ratio (m:f)	Patch size (ha)	Population size[a]	Density (animals/ha)
276b	1:4	9.0	5	0.56
490	1:1	1.6	2	1.25
C3	1:3	18.2	11	0.60
D3	1:1	8.3	2	0.24
E3	1:14	6.0	10	1.66

[a] Adult and subadult population size at beginning of radio-tracking.

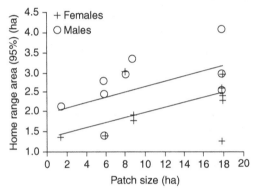

Figure 6.1. Relationships between home range size and patch size for each sex (redrawn from Pope *et al.*, 2004).

were apparent at Tumut, a radio-tracking study was undertaken of patch use by the arboreal marsupial, the Greater Glider. The work quantified home range size, den use behaviour and feed tree use in five patches of eucalypt forest of different sizes and shapes (Table 6.1) (Pope, 2003; Cunningham *et al.*, 2004b; Lindenmayer *et al.*, 2004b; Pope *et al.*, 2004).

Overall home range size (at 95th percentile isopleth level) in the Greater Glider decreased significantly with decreasing patch size and increased patch population density (Figure 6.1) (Pope *et al.*, 2004). Small patches had more animals per unit area, leading to smaller home ranges and greater home range overlap. Considerable overlap was observed between male and female gliders for the overall home range size. In contrast, home range core areas (50th isopleth level) remained relatively constant, regardless of patch size, population density or sex. This may indicate that core

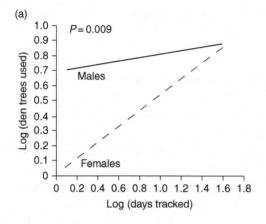

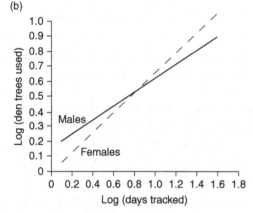

Figure 6.2. Den tree use by male and female Greater Glider (*Petauroides volans*): (a) Small eucalypt patches – for this graph, the patch size was set at 6 ha for illustrative purposes; (b) Large eucalypt patches – for this graph, the patch size was set at 18 ha for illustrative purposes (redrawn from Lindenmayer *et al.*, 2004b).

areas are an essential requirement for individuals and that the resources they contain cannot be shared with congeners (Pope *et al.*, 2004).

Radio-tracking work also involved tracking animals to their daytime den trees located in large hollow-bearing trees. The work showed that the number of different den trees used by individuals of the Greater Glider was lower in small patches than in large patches, especially among males (Figure 6.2) (Pope, 2003; Lindenmayer *et al.*, 2004b).

In contrast to the work on overall and core home range and den tree use, radio-tracking produced no evidence of a significant difference in patterns of behaviour between males and females of the Greater Glider in their night-time use of trees within remnant patches, regardless of patch size or population density (Cunningham *et al.*, 2004a).

Roosting behaviour by the Sulphur-crested Cockatoo in eucalypt patches

The Sulphur-crested Cockatoo is a large species of parrot native to Australia. Birds nest in tree hollows but also roost in various kinds of trees and other vegetation, often demonstrating long-term site affinity over many years (Lamm and Calaby, 1950).

Repeated spotlighting surveys at Tumut provided data on the roosting behaviour of the species in 40 patches of eucalypt forest of different size and shape and dominated by different tree species (Lindenmayer *et al.*, 1996). Analyses of data were conducted at the tree and patch levels and showed that:

- Birds almost always roosted in large Ribbon Gum (*Eucalyptus viminalis*) trees and only seven (of 173) records were from other tree species (Figure 6.3).
- The fewest roosting records were from the smallest patches, but patch size effects were non-linear because the highest number of observations of roosting were from intermediate-sized patches. This result was confounded by tree species, as most intermediate-sized patches were dominated by Ribbon Gum (Figure 6.3).

The reasons underpinning the variations in tree and patch-level roosting behaviour by the Sulphur-crested Cockatoo are not clear. They may, however, be associated with the suitability of different kinds of eucalypts as nest trees. An extensive survey completed at Tumut showed that Ribbon Gum eucalypts possess more hollows than other species of trees in the study area (Lindenmayer *et al.*, 2000c). It is possible that the apparent suitability of Ribbon Gum as a place to nest may also make them desirable roosting sites.

Bush Rat movements in remnant patches of eucalypt forest

A series of studies of the native Bush Rat using a combination of demographic and genetic approaches revealed some unexpected insights into the use of patches of remnant native forest by the species (Lindenmayer and Peakall, 2000; Peakall *et al.*, 2003, 2005; Lindenmayer *et al.*, 2005a; Peakall and Lindenmayer, 2006). The work involved catching animals in

(a)

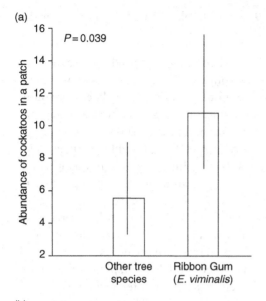

(b)

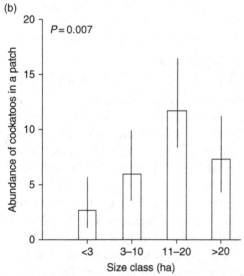

Figure 6.3. Relationships between the abundance of the Sulphur-crested Cockatoo at: (a) the tree species level; (b) the patch size level (redrawn from Lindenmayer *et al.*, 1996).

an array of eucalypt remnants of different size and isolation categories as well as in 'control' areas of continuous eucalypt forest. Estimates of population size were made, and tissue samples were gathered for subsequent genetic analyses. Further details of these studies are presented in the section on genetics (Chapter 9) in this book. Three key findings for the Bush Rat that are relevant to this chapter include:

- Bush Rat movements appeared to be restricted to patches connected by riparian vegetation along watercourses. Movements, and hence gene flow, between proximate patches not linked along gully lines were more restricted than predicted by simple measures of geographical (straight-line) distance between them (Lindenmayer and Peakall, 2000).
- Fine-scale patterns of genetic structure indicated highly restricted gene flow in populations of the Bush Rat within patches of remnant native vegetation. This was equivalent to a mean distance moved per generation of approximately 35 m (Peakall *et al.*, 2003, 2005).
- Restricted patterns of movement and gene flow within patches of remnant native vegetation were **not** an outcome of landscape modification. This was because similar patterns were quantified from contiguous areas of native forest where genetic and demographic studies of the Bush Rat were completed (Peakall *et al.*, 2003). Hence, limited movement and restricted gene flow was an innate condition in the species.

Bird calling behaviour within patches

Patterns of bird calling behaviour can be altered as a result of landscape change. For example, some authors have examined the process of 'conspecific attraction' (e.g. Reed and Dobson, 1993), whereby dispersing birds may be more likely to settle in habitat patches where there are already conspecifics vocalising (Smith and Peacock, 1990; Eens, 1994). This is because the amount of vocal activity is a cue indicating habitat suitability (Alatalo *et al.*, 1982).

A study at Tumut focused, in part, on relationships between landscape context and bird calling behaviour (Cunningham *et al.*, 2004b; Lindenmayer *et al.*, 2004a). Morning vocal activity data (or the 'dawn chorus') for birds were collected using automatic sound recorders at 165 sites in the cross-sectional landscape context study at Tumut (Figure 6.4). The recording device consisted of a tape-recorder housed within a metal ammunition box. A microphone was attached to a metal pole at 2 m above ground level. An insulated cable connected the microphone to the tape-recorder. The sensitivity of the microphone was 320 microvolts (µV) and 200–600 ohms (Ω). The recording system contained an automatic programmable switching system that enabled the device to be switched on or off for varying periods pre-set by the user.

Figure 6.4. Automatic call-box recording device seen here being calibrated against data gathered by an experienced ornithologist. Note that for most groups of birds, the observed relationships between vocal activity and the number of birds recorded by human point counts were statistically significant, but weak (Cunningham *et al.*, 2004b) (photo by David Lindenmayer).

Recording at each site took place in autumn 1997 and late spring 1997. A total of 1 hour of recordings was made across the morning at each site in each season. Recording began at 6.00 a.m. and the recording intervals selected were 6.00–6.10, 6.10–6.20, 6.20–6.30, 7.00–7.10, 8.00–8.10 and 9.00–9.10 a.m. on the same morning.

The call-box recording study demonstrated that vocal activity persisted longer at sites located within large areas of continuous eucalypt forest than in the strip- and patch-shaped eucalypt remnants surrounded by extensive stands of Radiata Pine or at sites dominated by stands of Radiata Pine. For remnants, vocal activity was greater in larger eucalypt patches and strips (i.e. >10 ha) than in smaller eucalypt patches and strips (Figure 6.5).

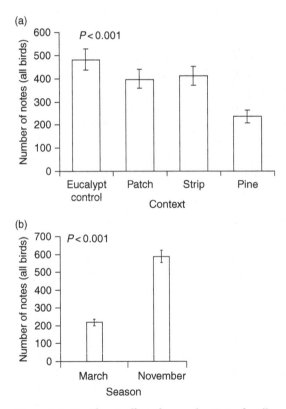

Figure 6.5. Significant effects for vocal activity for all species combined. Results are shown for: (a) Landscape context effect; (b) Season; (c) Mean longitudinal (hourly) vocal activity profiles for each landscape context; (d) Remnant area effect (redrawn from Lindenmayer *et al.*, 2004a).

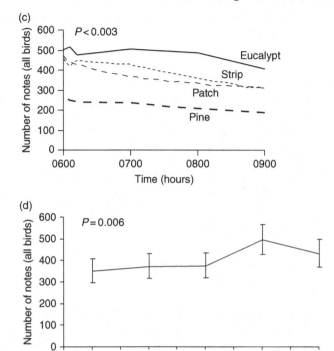

Figure 6.5. (cont.)

Vocal activity and landscape context relationships were identified not only for overall vocal richness but also for particular sets of species. Strong landscape context × season effects were found for understorey and under-canopy birds. Vocal activity for understorey and undercanopy birds was significantly lower in large areas of continuous eucalypt forest than the other landscape contexts, with the greatest levels of vocal activity being recorded in strip-shaped remnants.

There was a significant landscape context effect on vocal activity for honeyeaters, but the pattern among landscape context classes was not consistent between seasons (Figure 6.6). This effect was attributed to large differences in vocal activity between November and March 1997 for all landscape context classes except Radiata Pine, where vocal activity was lower in both seasons. There was evidence that temporal patterns (i.e. the longitudinal vocal activity profiles) were quite different depending on the type of site. For example, vocal activity for honeyeaters within large continuous areas of eucalypt forest first increased and then persisted until

(a)

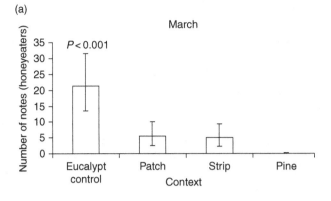

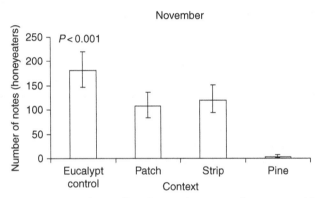

Figure 6.6. Significant effects for vocal activity in honeyeaters: (a) Landscape context × season effect; (b) Mean longitudinal vocal activity profile for each landscape context class; (c) Eucalypt remnant area effect (redrawn from Lindenmayer *et al.*, 2004a).

approximately 8.00 a.m. before declining, whereas in the eucalypt remnants it began declining after about 6.30 a.m. In addition, vocal activity for honeyeaters increased with size of remnant.

In summary, vocal activity for each bird group was almost always lower in the Radiata Pine sites than in sites in other landscape contexts. Thus, it seems that not only was bird abundance affected by factors related to landscape context (see Chapter 5), but also vocal activity. It is possible that the landscape-context differences in vocal activity might be associated with differences in the availability of food resources. Vocalising in birds has an energetic cost and the limited quantities of flower-based food resources (such as nectar, seeds and fruit) in Radiata Pine stands may limit the amount of surplus energy available to allocate to sound production or increase the amount of foraging time required to collect adequate

(b)

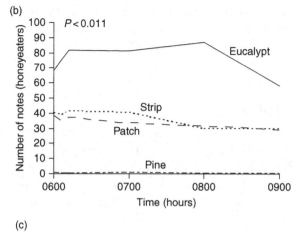

(c)

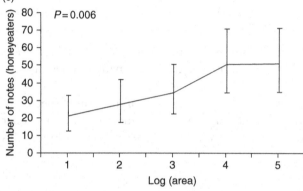

Figure 6.6. (cont.)

food resources. This explanation would be consistent with work by several authors who have demonstrated that vocal activity decreases with decreasing food resources (e.g. Davies and Lundberg, 1984; Cuthill and MacDonald, 1990).

Another key finding from the bird-call recording study was that vocal activity increased significantly with the size of eucalypt remnants. This result could be due to the increased resources available in larger patches – an outcome with a similar underlying cause to the Radiata Pine versus eucalypt forest context effect discussed above. However, the remnant size effect observed may also be associated with 'conspecific attraction', with the amount of vocal activity serving as a habitat suitability cue to other birds (Alatalo *et al.*, 1982). Therefore, increased activity in larger eucalypt remnants during the spring could be related to birds attempting to attract mates to colonise these areas. Whatever the basis for the observed results, they suggest that landscape patterns such as the one which gives rise to the mosaic of different forest types at Tumut (i.e. the different context classes) has an impact on vocal activity patterns in birds.

Patch–matrix interrelationships

Tubelis *et al.* (2004, 2007a) examined various aspects of remnant patch–pine matrix interrelationships for birds at Tumut. They surveyed birds at 32 sites that were boundaries between eucalypt patches and the surrounding pine plantation. Narrow (50–110 m wide) and wide (>300 m wide) remnant eucalypt patches were studied in which the adjacent pine stands were young (4–15 years old) or old (20–30 years old).

Tubelis *et al.* (2004) reported a range of novel and somewhat unexpected results. They found that:

- The use of the pine stands as additional foraging areas was most intense within 100 m of eucalypt patches for many species of birds (Figure 6.7).
- Bird species richness and the occurrence of particular species of birds decreased more rapidly in pine stands adjacent to wide rather than narrow patches of remnant native forest. They termed this

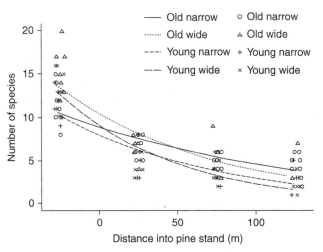

Figure 6.7. Mean bird species richness recorded at different distances from boundary lines in the four kinds of boundary treatment sites at Tumut (redrawn from Tubelis *et al.*, 2004).

phenomenon the 'halo effect', in which differential use of the matrix was significantly influenced by patch attributes. Tubelis *et al.* (2004) suggested that birds were unable to harvest sufficient food resources in narrow patches and compensated for this by foraging further into adjacent pine stands.

- The decrease in bird species richness from eucalypt patches was more pronounced when the surrounding pine stands were young (4–15 years old) than when they were 20–30 years old.

Subsequent studies (Tubelis *et al.*, 2007a, 2007b) showed that eucalypt patch width not only influenced overall species richness but also the occurrence of some kinds of birds, such as nectarivores and foliage searchers; the highest species richness and total bird abundance occurred in the widest patches.

In summary, the work by Tubelis *et al.* (2004, 2007a, 2007b) indicated that bird use of patches of remnant eucalypt forest is a function not only of patch width but also of the age of the surrounding pine stands. These various kinds of effects were not consistent across all groups of birds and some, such as those which forage by pouncing on their prey ('pouncers'; *sensu* Mac Nally, 1994), were more abundant in eucalypt patches surrounded by old-growth rather than young stands of pine (Tubelis *et al.*, 2007a). In addition, novel forms of use of the patch–matrix interface were identified, such as patch width effects on the distance into the surrounding pines ventured by some bird species (Tubelis *et al.*, 2004).

Altered breeding behaviour and dispersal

Studies of small mammal responses to landscape context revealed differences in occupancy between:

(1) Control sites and eucalypt remnants surrounded by stands of Radiata Pine;
(2) Eucalypt remnants of different size and spatial location (Lindenmayer *et al.*, 1999c; Chapter 5).

These results stimulated a number of additional studies aimed at isolating the mechanism(s) leading to such differences.

Banks *et al.* (2005a, 2005b, 2005c) studied the demography and genetics of the Agile Antechinus at Tumut. This multifaceted research programme examined patch occupancy, genetic variability, dispersal, breeding behaviour and a range of other factors in eucalypt control sites and eucalypt patches that ranged in size and isolation from areas of continuous eucalypt forest.

Large changes in the ecology of the Agile Antechinus were associated with landscape modification at Tumut. The landscape context of eucalypt patches (i.e. stands of Radiata Pine) was found to restrict dispersal, resulting in significantly lower gene flow (Banks et al., 2005a). Notably, barriers to dispersal (and hence gene flow) were not apparent in contiguous areas of native eucalypt forest. There were significant gender differences in the context effects on dispersal; it was restricted for males but not for females. This result was linked to the fact that female Agile Antechinus do not disperse – even in unmodified environments – and hence dispersal in the species is distinctly male-biased (Cockburn et al., 1985). Restrictions in male-biased dispersal were isolation-related and most pronounced in patches more than 750m away from large areas of contiguous native forest. Such restrictions did not occur in patches closer than 250m (Banks et al., 2005a). Isolated patches tended to be deficient in males and had smaller populations overall (Banks et al., 2005c). Thus, impaired dispersal appeared to be a key process governing the observed patterns of spatial variation in patch occupancy at Tumut (see Chapter 5).

A subsequent study by Banks et al. (2005b) further investigated the impacts of landscape context on the Agile Antechinus by quantifying aspects of breeding behaviour. Several important findings highlighted the way in which changes in patch use can manifest themselves in heavily modified environments. Restricted male dispersal (see above) resulted in elevated relatedness among potential mates in eucalypt remnants surrounded by pine stands. Communal nesting is a critical form of social interaction in the Agile Antechinus (Lazenby-Cohen, 1991), but rates of nest sharing among related members of the opposite sex were significantly higher in small patches than in large patches or in large areas of contiguous forest (Banks et al., 2005b). Thus, mechanisms for avoiding breeding with closely related individuals are impaired in human-modified landscapes where male dispersal is impeded. Other changes were quantified in the breeding behaviour of the species associated with landscape modification. For example, the amount of multiple paternity of litters was reduced in populations inhabiting small habitat patches in south-eastern Australia (Banks et al., 2005b).

Summary

This chapter describes a set of studies that have quantified how animals use patches of remnant native forest surrounded by extensive stands of Radiata

Pine. Different field methods (and sometimes laboratory techniques such as detailed genetic assays) were used to quantify various kinds of patch use such as movement patterns, bird vocalisation, dispersal and breeding behaviour. In many cases, this information contributed to an understanding of why and how landscape context and other effects of landscape modification have manifested themselves (see Chapter 5). Thus, some of the work has been an attempt to understand some of the ecological processes that have given rise to the patterns of patch occupancy and landscape context effects at Tumut.

It is notable that for some groups, such as small mammals, attempts to use models such as those used in population viability analysis (PVA) to predict patch occupancy proved to be particularly unsuccessful (see Chapter 8). A key problem was considered to stem from the fact that the ways in which species use contiguous forest may well be quite different from the ways in which they use eucalypt patches surrounded by pine stands (Lindenmayer et al., 2003a). New insights from detailed studies of patch use have helped to re-parameterise some of these models and, in turn, have improved the accuracy of predictions made using them (S.C. Banks et al., unpublished) – this topic is discussed further in Chapter 8.

7 · *Theory against data: testing ecological theories and concepts*

Ecology is characterised by a large and rapidly increasing body of theory. Empirical data are the best tests of ecological theory (Franklin and MacMahon, 2000; Krebs, 2008). Such tests are critical, especially if ecological theory is eventually to become useful in applied natural resource management, including the conservation of biodiversity (Doak and Mills, 1994; Fazey *et al.*, 2005).

This chapter explores some of the tests of ecological theory and ecological concepts completed using datasets from Tumut. First, some results are summarised that indirectly examined the results of studies of landscape context effects (see Chapter 5) and their relationships with different conceptual models of landscape cover (see Chapter 2 for a discussion of these kinds of models). The second section of this chapter is a précis of research on nested subset theory. Third, the results of work on thresholds in the amount of native vegetation cover are outlined. Research on landscape indices is summarised in the fourth section of this chapter. The final section focuses on evidence for the peninsula effect at Tumut.

Each section of this chapter comprises a short overview of the theory or the ecological concept being examined, a summary of research findings and a brief discussion of some of the lessons that emerged from the work. In some cases, to facilitate tests of a particular ecological theory or concept, new datasets had to be gathered to complement information that had already been collected. A short description of the survey methods and field data is presented in these instances.

Conceptual models of landscape cover

As outlined in Chapter 2, many conceptual landscape models have been proposed over the past few decades to help characterise landscapes through:

(1) Grouping landscape elements into classes;
(2) Allocating entire landscapes to a particular class based on the amount and distribution of landscape elements;

(3) Recognising the multiple perspectives of landscapes by different organisms (Lindenmayer and Hobbs, 2007).

These include such conceptual landscape models as:

- *The island model* (Shafer, 1990), which has its origins in island biogeography theory (MacArthur and Wilson, 1963, 1967);
- *The patch–corridor–matrix model* (Forman, 1995);
- *The hierarchical patch dynamics model* (Wu and Loucks, 1995);
- *The landscape variegation model* (McIntyre and Barrett, 1992);
- *Species-specific gradient models* (Austin, 1999; Fischer *et al.*, 2004; Manning *et al.*, 2004).

These different conceptual landscape models can be used to classify, study and understand landscapes on the basis of different entities such as:

- Structural attributes such as the amount and configuration of vegetation (e.g. Forman, 1995; McIntyre and Hobbs, 1999);
- Habitat for a particular species (e.g. Fischer *et al.*, 2004; Manning *et al.*, 2004);
- Functional attributes or landscape processes (e.g. Ludwig *et al.*, 1997).

Empirical studies of conceptual landscape models at Tumut

Fischer and Lindenmayer (2006) summarised the findings for the various biotic groups and individual species that were studied at Tumut and examined them in the context of the continuum model of landscape cover and the island and patch–corridor–matrix models more typically used in studies of landscape change and habitat fragmentation. First, they outlined the key assumptions of the two broad kinds of models (Table 7.1). They then outlined which results for particular groups conformed best to a given conceptual model of landscape cover (Fischer and Lindenmayer, 2006).

Table 7.2 is a broad overview of some of the findings at Tumut for different vertebrate groups and particular species of vertebrates. This synthesis clearly shows that the findings for some species conformed best to the island/patch–corridor–matrix model, whereas for others, the continuum model was the most appropriate conceptual model of landscape cover.

Fischer and Lindenmayer (2006) noted that the responses of several species were not congruent with the assumptions that underpin the island/patch–corridor–matrix model. In particular:

Table 7.1. *Assumptions underpinning the island/patch–corridor–matrix models versus the continuum model of landscape cover (modified from Fischer and Lindenmayer, 2006).*

	Island/patch–corridor–matrix models	Continuum model
Landscape pattern	Assumes clear contrast between patches and areas outside patches	Allows landscape with gradually changing patterns
The notion of 'patches'	Based on human-defined patch boundaries to correspond closely with animal-perceived patch boundaries; patches are assumed to be internally homogeneous	Human-defined patches are not of primary interest; no assumptions are made about internal homogeneity
Identity of species	Restricted to single species or multiple species with similar requirements	Allows consideration of multiple species with marked differences in requirements
Species distributions	Requires species to be restricted to patches, ideally as metapopulations	Allows species distribution in space in complex and sometimes continuous ways
Ecological processes	Assumes that landscape pattern is a reasonable proxy for many interacting ecological processes	Attempts to study ecological processes directly

- Some species were not confined to human-defined 'habitat patches';
- Species within a particular group varied markedly in their response to the pattern of vegetation cover;
- Food, shelter and climatic variables were often as important as landscape pattern in their influence on the distribution and abundance of biota.

Lessons from the research at Tumut

The vast majority of ecologists remain unaware of some conceptual landscape models, such as the continuum model, the landscape variegation model or the hierarchical patch dynamics model (Lindenmayer and Hobbs, 2007). Many ecologists have traditionally employed (often unwittingly) the island model or the patch–corridor–matrix model to classify landscapes and guide their research or management, particularly in landscapes subject to human modification (Haila, 2002). Based on the extensive work at Tumut, it is clear that different models have their own strengths and limitations and that no one model will perform best in all

Table 7.2. *Summary of findings for different vertebrate groups and individual species as they relate to the island/patch–corridor–matrix model or the continuum model of landscape cover (modified from Fischer and Lindenmayer, 2006).*

Vertebrate group	Finding(s)
Fragmentation-related findings	
Birds	Distribution patterns of Sacred Kingfisher (*Todiramphus sanctus*) predicted well by metapopulation model (Lindenmayer *et al.*, 2001c)
	Bird species richness highest in large patches and continuous native forest (Lindenmayer *et al.*, 2002a)
Marsupials	Greater Glider (*Petaurus volans*) and Common Brushtail Possum (*Trichosurus vulpecula*) more likely in large patches (Lindenmayer *et al.*, 1999a)
	Yellow-bellied Glider (*Petaurus australis*) only in continuous native forest (Lindenmayer *et al.*, 1999a)
	Brown Antechinus (*Antechinus stuartii*) more likely in well-connected patches (Lindenmayer *et al.*, 1999c)
Rodents	Bush Rat (*Rattus fuscipes*) may use native streamside vegetation as corridors (Lindenmayer and Peakall, 2000)
	Bush Rat more likely in large patches (Lindenmayer *et al.*, 1999c)
Lizards	Maccoy's Skink (*Nannoscincus maccoyi*) more likely in less isolated patches (Fischer *et al.*, 2005)
	Coventry's Skink (*Niveoscinus coventryi*) more likely in large patches (Fischer *et al.*, 2005)
	Lizard species richness highest where context was dominated by eucalypt forest (Fischer *et al.*, 2005)
Frogs	Only two frog species in the pine matrix (Parris and Lindenmayer, 2004)
Findings better explained by the continuum model	
Birds	Laughing Kookaburra (*Dacelo novaeguineae*) may move frequently between patches (Lindenmayer *et al.*, 2001c)
	Predefined patch boundaries may not apply to the White-throated Treecreeper (*Cormobates leucophaea*) (McCarthy *et al.*, 2000)
	Different bird species perceived the same landscape as fragmented, variegated or continuous (Lindenmayer *et al.*, 2003b)
Marsupials	Common Ringtail Possum (*Pseudocheirus peregrinus*) more likely in forest patches than in continuous native forest (Lindenmayer *et al.*, 1999a)
	Common Wombat (*Vombatus ursinus*) and Swamp Wallaby (*Wallabia bicolor*) regularly in pine forest (Lindenmayer *et al.*, 1999c)
	Arboreal marsupials feeding on insects, pollen, plant and animal exudates more common in continuous forest (i.e. explicit link with food) (Lindenmayer *et al.*, 1999a)

Table 7.2. (*cont.*)

Vertebrate group	Finding(s)
Rodents	Bush Rat may use streamside vegetation for dispersal because it offers shelter and wet microclimate (Lindenmayer and Peakall, 2000)
Lizards	Lizard occurrence related to invertebrate abundance (i.e. food availability) (Fischer et al., 2005)
	Different lizard species had different altitudinal preferences, presumably in response to climate (Fischer and Lindenmayer, 2005c)
	Garden Skink (*Lampropholis guichenoti*) and Coventry's Skink more abundant at plots with a high volume of old logs to shelter under (i.e. patches were not homogeneous) (Fischer et al., 2005)
	Distribution patterns of several lizard species extended into the pine forest (Fischer et al., 2005)
	Maccoy's Skink avoided canopy gaps, presumably in response to microclimate (i.e. patches were not homogeneous) (Fischer et al., 2005)
	Garden Skink highly abundant at some recently clear-cut pine sites (Fischer et al., 2005)
Frogs	Frogs responded to moisture regime and vegetation structure, rather than patch size (Parris and Lindenmayer, 2004)

circumstances (Fischer and Lindenmayer, 2006). There are at least six other reasons why a pluralistic approach to conceptual models of landscape models is important. These are (after Lindenmayer et al., 2007a):

- Every landscape is complex;
- Every landscape is different;
- No landscape is static;
- Different elements of the biota perceive the same landscape differently;
- Landscapes embody both patterns and processes, and these are multi-scaled entities;
- Many landscape patterns and processes are continuous entities or gradients rather than discrete entities marked by sharp boundaries.

Different problems, objectives and goals may require different landscape models and different landscape classifications. For example, a classification and a model suitable to guide a particular research programme may be markedly different from one needed to address the practical problems faced by a landscape manager (Lindenmayer and Fischer, 2006). A key question should therefore not be: 'Is a patch model, a continuum model or some other model better for classifying landscapes or describing landscape use by fauna?' Instead, a more apt question is: 'In which situations is a patch model appropriate, and in which situations does a continuum model or some other kind of conceptual landscape model work better?'

Nested subset theory

As discussed in Chapter 2, an assemblage distributed across a number of discrete sites is considered nested when the taxa present at relatively species-poor sites are subsets of those present at progressively more species-rich sites. Nestedness analysis is a way of examining both the species richness and the composition of a particular assemblage.

Empirical studies of nestedness at Tumut

A range of the datasets gathered at Tumut were appropriate for testing whether patterns of animal occurrence in different patches of remnant native vegetation were nested. In particular, complete patch counts of arboreal marsupials had been completed on a repeated basis for many sites (see Chapter 4). For birds, complete patch counts, transect counts and automatic bird call recording (see Chapters 4, 5 and 6) were undertaken, and composite data compiled for sites. Data gathered for reptiles at Tumut (Fischer *et al.*, 2005) were also appropriate for nested subset analysis.

The arboreal marsupial assemblage at Tumut comprised six species (Chapter 5). Detailed analyses indicated that it was significantly non-nested (Table 7.3). Some interesting patterns of co-occurrence were identified. For example, the Greater Glider and the Common Ringtail Possum almost never co-occupied patches of remnant native forest, whereas co-occupancy was five times more prevalent in sites within contiguous areas of native forest (Fischer and Lindenmayer, 2005a). Such species-specific differences in response to landscape context (see Chapter 5), coupled with the limited number of species, are likely to be the major reason for the lack of significant nestedness observed for the arboreal marsupial assemblage at Tumut (Fischer and Lindenmayer, 2005a).

The assemblage of lizards at Tumut comprised 14 species and was also significantly non-nested (Fischer and Lindenmayer, 2005a; Table 7.3). The main reason for the lack of nestedness was that the pool of potentially occurring species was not the same at all sites – high-elevation sites supported a different suite of lizard species than low-elevation sites (Fischer and Lindenmayer, 2005c).

Table 7.3. *Global tests of nestedness for birds, arboreal marsupials and reptiles at Tumut (based on Fischer and Lindenmayer, 2005a).*

Group and dataset	Number of species	Number of sites	Sites sorted by species richness	Sites sorted by area	Sites sorted by isolation
Birds					
Pine sites	69	40	$P=0.047$	–	–
Contiguous forest	75	40	$P=0.118$	–	–
Remnant patches	76	43	$P<0.005$	$P<0.005$	$P=0.611$
Arboreal marsupials					
Contiguous forest	7	31	$P=0.115$	–	–
Remnant patches	6	33	$P=0.817$	$P>0.995$	$P>0.995$
Reptiles	13	30	$P=0.117$	$P=0.685$	$P=0.953$

The dataset for birds was far more extensive and species-rich than those for arboreal marsupials and reptiles. It was therefore suitable for examining a wider range of questions than for the other two vertebrate groups. Fischer and Lindenmayer (2005a) found that the bird assemblage at Tumut was significantly nested. Significant nestedness also characterised many of the subsets of the data that were examined (Table 7.3). For example, nestedness was found for birds at sites in Radiata Pine stands and also those in patches of remnant native forest. In contrast, populations of birds in sites within continuous areas of native forest were not nested (Fischer and Lindenmayer, 2005a). Although the bird populations in the first two kinds of sites exhibited highly significant nestedness, a large amount of variation characterised the datasets (Figure 7.1).

When birds in patches of remnant native forest were examined, highly significant nestedness was also found for patch area, but not for patch isolation (Table 7.3). Further subdivision of the dataset for birds involved identifying those species thought to be potentially sensitive to human landscape modification (e.g. small species with low mobility and habitat specialisation) and segregating sites with high and low levels of environmental heterogeneity, as reflected by the occurrence of particular forest types. Such data partitioning typically strengthened nestedness relationships for birds (Fischer and Lindenmayer, 2005a).

Lindenmayer and Fischer (2006) noted that three ecological mechanisms are typically considered to be particularly important with regard to giving rise to nestedness. These are selective extinction, selective immigration and habitat nestedness. Data analyses at Tumut indicated that

43 patches sorted by species richness

76 species sorted by number of occurrences

Figure 7.1. Species × site matrix for birds in patches of remnant native forest surrounded by pine stands at Tumut. The figure shows a high degree of scatter in the data, despite the existence of highly significant nestedness (redrawn from Fischer and Lindenmayer, 2005a).

patch isolation was not significant for birds (Table 7.3), suggesting that selective immigration was unlikely to be a key factor driving the observed patterns for birds. However, finding a useful common metric for isolation that is relevant to all species of birds in an assemblage is far from straight-forward. The lack of significant nestedness for sites in contiguous areas of native forest suggested that habitat nestedness was unlikely to be an important driver underpinning the patterns for birds identified at Tumut (Fischer and Lindenmayer, 2005a). Analyses relating nestedness to the area of remnant eucalypt patches (Table 7.3) were strongly suggestive of selective extinction being a primary mechanism for nestedness. The trend for strengthened nestedness for bird assemblages dominated by habitat specialists and relatively immobile species was consistent with that conclusion (Fischer and Lindenmayer, 2005a). Thus, the suite of analyses of nestedness helped to identify some of the mechanisms that have given rise to the patterns of bird species composition that characterised the study area.

Lessons from the research at Tumut

Like island biogeography theory (MacArthur and Wilson, 1967), nested subset theory has also been used to examine issues such as patterns of species distribution in human-modified patchy landscapes (Doak and Mills, 1994). Because nestedness analyses are a way of studying species composition as well as species richness, nested subset theory has also received attention as a potential tool to investigate whether a single large reserve or several small ones are preferable for species conservation (Patterson, 1987; Cutler, 1991; Boecklen, 1997). Most recently, nestedness has been suggested as a tool for picking surrogate species in an attempt to provide a scientifically defensible shortcut for species conservation (Smyth et al., 2000; but see Fleishman et al., 2002).

In human-modified landscapes, nestedness analyses can be useful for providing an overview of likely conservation issues. For example, they can show whether local immigration or extinction are likely drivers of community composition, and if large or well-connected patches are particularly important for conservation. However, given the community-level nature of nestedness tests, statistically significant nestedness should be interpreted with caution before being used to develop conservation guidelines (e.g. regarding what size patches to protect; see Patterson, 1987; Rosenblatt et al., 1999; Berglund and Jonsson, 2003). Particular caution is warranted when using nested subset theory as a means of identifying ecological indicators (see Honnay et al., 1999; Smyth et al., 2000) because important species-specific differences and idiosyncrasies may be masked by global tests of nestedness (Simberloff and Martin, 1991). Given this, the identification of nestedness does not necessarily lead to clear insights for maintaining subsets of species over time (Whittaker, 1998). On this basis, Fischer and Lindenmayer (2005a) recommended that nestedness analyses should be complemented by detailed analyses at other levels of biological organisation, particularly individual species of conservation concern.

Box 7.1.   The conservation implications of significant nestedness versus perfect nestedness

Fischer and Lindenmayer (2005b) investigated the implications of imperfect nestedness in a conservation context in some detail. They compiled presence/absence data on birds in 43 native eucalypt forest patches scattered throughout an exotic softwood plantation in south-eastern Australia. The bird assemblage was significantly nested when patches were sorted by their size. However, there were some 'holes' and some 'outliers', i.e. unexpected absences and presences of birds at some sites.

If nestedness had been perfect – i.e. if there had been no holes and no outliers – the single largest patch would have contained all bird species, including those of conservation concern. Arguably, perfect nestedness would therefore have indicated that the largest patch was disproportionately important for bird conservation. However, because the dataset was significantly (but not perfectly) nested, substantially more sites than the single largest patch alone were needed to capture most bird species. For example, fewer than half of the bird species classified as 'sensitive to landscape change' co-occurred in any given patch. Using the largest patches only, more than a quarter of patches surveyed were required to capture 80% of sensitive species in at least one patch (Fischer and Lindenmayer, 2005b).

Small patches have many limitations, such as increased susceptibility to edge effects (Lovejoy *et al.*, 1986) and limited food resources (Zanette *et al.*, 2000). However, even when an assemblage is significantly nested in relation to patch size, small patches can make an important complementary contribution to large patches and should not be neglected in conservation management.

Probably the most important caveat of nestedness analyses in modified landscapes is that they are based on the island model (see above). Nestedness analyses in modified landscapes sometimes (implicitly) consider patches of native vegetation as spatially discrete 'islands' surrounded by 'unsuitable land' (Doak and Mills, 1994). This view can be overly pessimistic (see Chapter 5) and may fail to recognise the potential value of conservation strategies in the areas surrounding vegetation remnants for at least some of the species in an assemblage (Lindenmayer and Fischer, 2006).

Ecological thresholds in the amounts of native vegetation cover

Chapter 2 described some of the background associated with ecological threshold theory and its relationships with the decline in biota and/or the deterioration of key ecological processes. In particular, it has been speculated that cover levels of 30% in native vegetation might be critical change points or thresholds for biodiversity in landscapes. That is, the loss of species and populations below the 30% 'threshold' will be greater than can be predicted from a smooth relationship with vegetation cover alone (Figure 7.2) (Huggett, 2005; Chapter 2).

Empirical studies of thresholds at Tumut

Data gathered at Tumut on birds and reptiles were used in statistical analyses that attempted to identify threshold relationships between the amount of native vegetation cover and species diversity and occurrence of individual bird and lizard taxa (Lindenmayer *et al.*, 2005b). For birds, the data used were the point-interval counts from the randomised selection of 86 eucalypt remnants in the study area (see Chapter 3); these remnants varied in size from 1 to 124 ha. In the case of lizards, the logistics of pitfall trapping meant that fewer locations could be sampled than for birds. Altogether, 30 sites were sampled, including 24 eucalypt sites (ranging from 1 ha in size to continuous forest), three sites in pine forest >20 years of age and three sites in recently clear-cut pine forest. Statistical analysis involved fitting generalised additive models (see Hastie and Tibshirani, 1990) and 'broken-stick' models (two intersecting lines of different slope).

No strong empirical evidence was found for a threshold relationship (e.g. a 'broken-stick' model; see Figure 7.3) between bird species richness

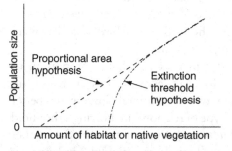

Figure 7.2. Hypothesised vegetation cover threshold for species richness (redrawn from Fahrig, 2003).

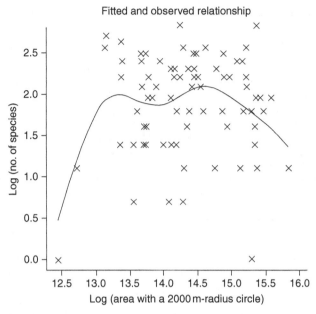

Figure 7.3. Relationship between bird species richness and the natural logarithm of the area of native vegetation within a circle of 2,000 m radius. The solid line is the fitted relationship estimated using a general non-parametric spline smoother, which was not statistically significant (see Hastie and Tibshirani, 1990). Redrawn from Lindenmayer *et al.* (2005b).

(excluding pine-using species) and the total area of native vegetation within a circle of 2,000 m radius around the 86 sites (Figure 7.3). There was also insufficient signal in the data to distinguish between many alternative plausible models (such as straight-line and curvilinear functions) (Lindenmayer *et al.*, 2005b). Furthermore, there was no empirical evidence for a change-point or threshold relationship between the probability of detection of individual bird species and the area of native vegetation cover. Findings for the Red Wattlebird (*Anthochaera carunculata*) (Figure 7.4) were indicative of those for other bird taxa that were analysed (Lindenmayer *et al.*, 2005b).

Lizard species richness significantly increased with the proportion of native vegetation within 1,000 m. However, as with birds, there was no evidence that a threshold relationship described this relationship any better than a smooth, continuous or other type of relationship. Forcing a threshold relationship onto the data resulted in two change points in the data, rather than one. Neither change point was ecologically meaningful (Lindenmayer *et al.*, 2005b).

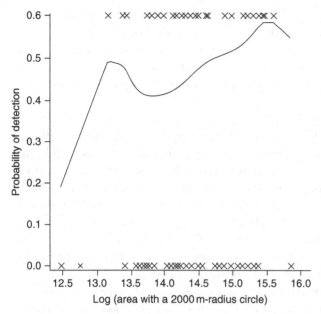

Figure 7.4. The probability of occurrence of the Red Wattlebird and the natural logarithm of the amount of native vegetation within a surrounding landscape area of 2,000 m. The solid line is the fitted relationship estimated using a general non-parametric spline smoother, which was not statistically significant (see Hastie and Tibshirani, 1990).

Lessons from the research at Tumut

The existence of a simple threshold amount of native vegetation below which ecosystems undergo regime shifts is appealing from both a scientific and a management perspective. However, several important caveats of the notion of such thresholds should be considered, particularly in the context of a simple threshold postulated to apply to many landscapes, such as the '30% rule'. First, a hypothesised threshold amount of native vegetation below which species richness declines more rapidly hides the fact that species become extinct with native vegetation loss even *above* a postulated threshold level – albeit at a slower rate.

Second, a hypothesised threshold does not consider that land clearing often takes place primarily in areas with high primary productivity; therefore the remaining amount of native vegetation is rarely a representative sample of the original land cover.

Third, with respect to species richness of assemblages, it is important to note that the '30% rule' is based on a binary classification of land into patches of native vegetation that are deemed suitable for all species in a given assemblage versus a matrix of universally unsuitable land. While this classification may be a reasonable approximation of assemblage organisation in some landscapes, often the matrix and non-native vegetation and structural features within it can play important ecological roles (see above and Chapter 2).

Fourth, not all species respond to a given pattern of native vegetation in the same way (Mönkkönen and Reunanen, 1999). Therefore, different levels of native vegetation may be suitable for different species (Homan et al., 2004), and some species may have continuous or linear rather than threshold relationships with native vegetation cover (Bunnell et al., 2003). For species richness – a key characteristic of ecosystems which emerges from many species' individualistic responses – this means that its particular relationship with native vegetation will be highly situation-specific and will vary between taxonomic groups and landscapes (see Parker and Mac Nally, 2002; Drinnan, 2005; Radford et al., 2005).

Thus, although there are cases where regime shifts have been associated with the '30% rule' (e.g. Andrén, 1994), the four key considerations discussed above illustrate that the rule (like all ecological principles) cannot be applied uncritically in all landscapes, ecosystems and assemblages (Groffman et al., 2006). Perhaps the most useful, albeit general, conclusion from the literature on vegetation thresholds to date is that cascades of negative effects on native species assemblages arising from the simultaneous losses of native vegetation and landscape connectivity are more likely to occur at low levels of native vegetation cover.

Despite the issues outlined above, there are cases where threshold responses to native vegetation cover or the amount of available habitat have been identified (Sasaki et al., 2008). The best examples of such analyses have involved targeted tests of individual species for which precise habitat requirements are known and where the amount of available habitat is accurately quantified. Homan et al. (2004) provide an elegant empirical example based on work on salamanders in North America.

In the case of the work at Tumut, additional studies of threshold effects in landscape cover might focus on efforts to define precisely the habitat requirements of particular species (a difficult task; see Burgman et al., 2005; Morrison et al., 2006), map habitat availability at appropriate spatial scales for those individual taxa, and then recommence searches for threshold effects for them.

Landscape indices

Spatial patterns of vegetation cover created by landscape alteration are often linked with the responses of particular elements of the biota. Metrics to characterise such patterns are termed landscape indices (Wegner, 1994; Forman, 1995; Smith, 2000) and are widely used in studies of modified landscapes, particularly in observational studies (Haines-Young and Chopping, 1996; Gustafson, 1998; McAlpine *et al.*, 2002b).

Landscape indices attempt to provide descriptions of spatial landscape patterns, particularly 'patchiness', and the size, shape, composition, juxtaposition and arrangement of landscape units (e.g. vegetation patches). Such indices typically attempt to capture three broad groups of phenomena:

- *Composition*, or the identity and characteristics of the different types of patches;
- *Configuration*, the spatial arrangement of patches and temporal relationships between these units;
- *Connectivity*, either from a human perspective (landscape connectivity; see Chapter 12) or from a species-specific perspective (habitat connectivity; see Chapter 6).

There are vast numbers of landscape indices (Turner, 1989; Wegner, 1994; reviewed by Haines-Young and Chopping, 1996) and their use in studies around the world is increasing (Morgan, 2001). Landscape indices may be valuable to help characterise a landscape and establish whether a pattern has changed (Wegner, 1994; Smith, 2000; McAlpine *et al.*, 2002b). They may also be valuable for linking spatial patterns of landscape cover to species responses. However, when using landscape indices to describe landscapes and guide management actions, it is important to be clear about the reason for using them and the inherent limitations of using such an approach (Tischendorf and Fahrig, 2000a, 2000b).

Empirical studies of landscape indices at Tumut

Studies of landscape indices at Tumut have focused on data gathered from the 86 eucalypt remnants identified for the landscape context study of birds and mammals (see Chapter 5). For each remnant, 'landscape areas' were defined by concentric circles of radius 200 m, 400 m and 2,000 m around the centroid of each of the 86 eucalypt remnants. Data relating to several landscape measures (see below) were derived using a geographical information system (GIS). Surrogate measures were chosen on the basis

Table 7.4. *Landscape indices used at Tumut.*

Code for landscape measure	Description
REMNANT AREA	The precise size (to the nearest 0.1 ha) of a given eucalypt remnant within which the survey site was located
AREA	The total area of native vegetation cover (m^2) within the 'landscape area' targeted for study (i.e. the circle of radius 200 m, 400 m, 2,000 m around the site centroid)
NP	Number of native eucalypt patches within the study polygon
MPS	Mean area of native vegetation patches within the study polygon; a measure of spatial subdivision
MPSSTD	Standard deviation of MPS values within the study polygon; a measure of variability in patch size
LPI	Largest patch index, which was the proportion (%) of the study polygon covered by the largest patch of native vegetation; a measure of spatial subdivision
TCA	Total core area, which was the sum of the area of 'core' native vegetation. The 'core' area is the area of a given patch that is at least two cell widths (40 m) from the patch edge; an integrated measure of patch edge and shape
NMEAN	Mean minimum distance (m) to native vegetation (up to a maximum of 200 m) of all cells within the study polygon; a measure of isolation
RDEN	Density of roads within the study polygon measured as the sum of all cells within the study polygon bisected by a road

of their relative computational and intuitive simplicity; and were as follows (see also Table 7.4):

- The amount of (eucalypt forest) habitat in a landscape;
- The spatial subdivision of habitat in a landscape;
- Patch isolation;
- Patch shape and edge (or core) area;
- Relative density of roads in the landscape area surrounding each of the 86 eucalypt remnants.

The size and shape of many eucalypt remnants meant that the 200 m and 400 m polygons frequently encompassed only the remnant itself and so were not large enough to capture information on the characteristics of the landscape surrounding it. Hence, detailed statistical analysis was restricted to data from the 2,000 m polygon.

Given the compositional nature of the variables, collinearity among the variables was not unexpected. To meet linearity and distributional assumptions, all variables were transformed by natural logarithms prior to statistical analysis. Dimension reduction by principal components analysis enabled the identification of a set of vectors or scores. The first score was a contrast between remnant size and the mean area of native eucalypt forest within the 2,000 m polygon versus the number of eucalypt remnants and the mean minimum distance to other areas of native vegetation within the 2,000 m polygon. Hence, the contrast essentially reflected whether the eucalypt vegetation was consolidated (a negative score) or patchy (a positive score).

The following target response variables were used for exploring relationships with various landscape surrogates:

- The total number of arboreal marsupials recorded;
- The presence/absence of the Greater Glider and the Common Ringtail Possum;
- Detection rates for the following bird species: Crimson Rosella, Grey Fantail, Red Wattlebird, Eastern Yellow Robin, Golden Whistler and Rufous Whistler.

Regression analysis showed significant relationships between the occurrence of most species examined and landscape variables. For example, there was evidence that the probability of detecting a Greater Glider in a patch increased as the log of remnant area increased in the surrounding landscape. For the Common Ringtail Possum, the probability of occurrence of the species increased significantly with the log (mean number of remnants), but decreased significantly with log(total core area).

Regression analysis also showed significant relationships between the detection frequency of birds and landscape variables (Figure 7.5). Each relationship contained a different set of significant explanatory variables, even for the two species from the same genus (the Rufous Whistler and Golden Whistler). In other cases, a surrogate measure was significant in models for two species, but the nature of the response was different. For example, there was a positive relationship between the probability of detection of the Grey Fantail and mean minimum distance to native vegetation, but a negative one for the Golden Whistler. For the Crimson Rosella, the probability of detection increased significantly with log(total core area), but decreased significantly with log(mean area of remnants) and log(area of native vegetation).

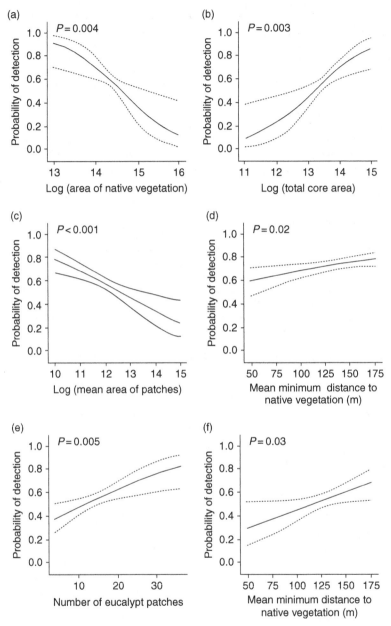

Figure 7.5. Relationships between the detection frequencies of various species of birds and significant landscape variables: Crimson Rosella (a-c), Grey Fantail (d), Rufous Whistler (e-g) and Red Wattlebird (h-k) (redrawn from Lindenmayer *et al.*, 2002b).

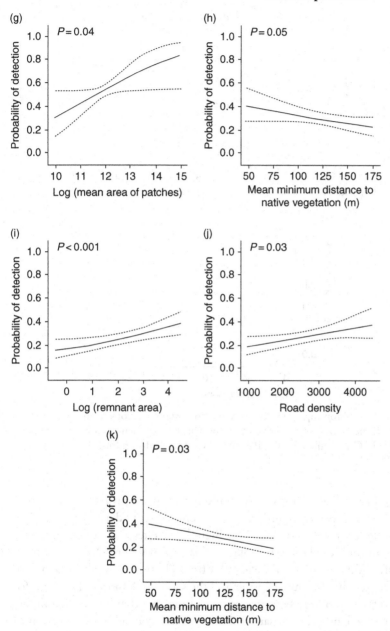

Figure 7.5. (cont.)

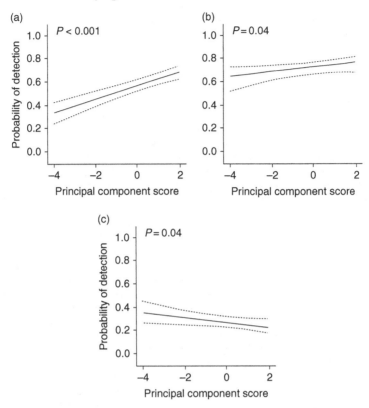

Figure 7.6. Relationships between the detection frequencies of various species of birds and a patchiness score derived from principal components analysis: Crimson Rosella (a), Grey Fantail (b) and Red Wattlebird (c) (redrawn from Lindenmayer *et al.*, 2002b).

Significant relationships were found between the probability of detection of several species of birds and the patchiness score derived by principal components analysis (Figure 7.6). Significant effects included those for the Crimson Rosella, Grey Fantail and Red Wattlebird. No effects were identified for the Eastern Yellow Robin. There was weak evidence of an effect for the Rufous Whistler and the Golden Whistler. The signs for the effects varied between species, indicating that they responded differently to the landscape mosaic at Tumut. The probability of detection of the Red Wattlebird and Golden Whistler was greater if remnant eucalypt cover was consolidated, whereas for the Crimson Rosella, Grey Fantail and Rufous Whistler it was greater if remnant eucalypt cover was patchy and dispersed among many remnants.

Lessons from the research at Tumut

Different combinations of landscape surrogate measures were found to predict the presence/abundance of different species. This result was not surprising. In models with the same significant explanatory variable, the sign for trends in the variable was often different (e.g. for the mean minimum distance to native vegetation for the Grey Fantail and the Golden Whistler). Hence, each species responded differently to the land-scape mosaic at Tumut – a result consistent with other studies where the response of an array of taxa to landscape modification has been examined (e.g. Robinson *et al.*, 1992; Gascon *et al.*, 1999; Barlow *et al.*, 2007).

The study of landscape indices at Tumut produced several unexpected findings. For example, the abundance of arboreal marsupials increased significantly with remnant area, but decreased with an increasing amount of native vegetation in the surrounding landscape. Thus, 'isolated' rem-nants surrounded predominantly by Radiata Pine supported significantly more animals than did those of equivalent size but where the surrounding landscape included other areas of native vegetation. Hence, population dynamics within a remnant appears to be influenced by what is occurring outside it (in this case the surrounding Radiata Pine stands). A 'fence effect' (*sensu* Wolff *et al.*, 1997) may be one explanation for this outcome; the unsuitability of the surrounding Radiata Pine plantation may make animals reluctant to disperse away from eucalypt remnants. This would increase population density relative to those remnants where other areas of eucalypt forest occurred in the surrounding landscape. This type of effect is not often reported, but has been seen in small mammals in heavily modified landscapes in the Northern Hemisphere (e.g. Krohne, 1997; Bayne and Hobson, 1998). Fence effects demand some plasticity in life history parameters, such as increased population density, increased home range overlap and reduced core home range size, as a consequence of remnant conditions. Notably, radio-tracking of the Greater Glider at Tumut showed that home range size changes with remnant size (Pope *et al.*, 2004; see Chapter 6).

Work on landscape indices demonstrated that many species respond not only to the total area of eucalypt cover, but also to how that cover is spatially arranged. Some taxa were more likely to be recorded where eucalypt cover was consolidated (e.g. Red Wattlebird), whereas the opposite effect was identified for other species (i.e. eucalypt cover was dispersed – e.g. Crimson Rosella). Similarly, the Common Ringtail Possum was more likely to occur in a patchy (rather than consolidated) distribution of eucalypt remnants. This effect may be due to positive responses to edge conditions. The eucalypt–pine boundary may provide additional resources for some species. For example, the Common Ringtail Possum can use eucalypt forests and neighbouring stands for both food and shelter (Friend, 1980). A combination of forest types having shared boundaries may provide more resources than either Radiata Pine stands or eucalypt forest alone. These positive edge effects may explain the results of the earlier studies of arboreal marsupials (Lindenmayer *et al.*, 1999a; see Chapter 5) in which there was a significantly higher abundance of the Common Ringtail Possum in eucalypt remnants (surrounded by Radiata Pine) than in large continuous stands of eucalypt forest.

The work at Tumut produced a number of more general perspectives on landscape indices. For example, it was noted that:

- The methods used to develop indices are often not provided.
- It is not always clear what to do with landscape indices once they are generated because they do not necessarily link closely to land management.
- Few measures are used consistently in different investigations.
- Landscape indices generally fail to account for factors such as the vertical complexity of vegetation, which is known to be important for bird species richness and abundance (e.g. MacArthur and MacArthur, 1961; Gilmore, 1985; Brokaw and Lent, 1999).
- Many indices provide sophisticated ways of highlighting intuitively obvious landscape patterns and have led to few new insights.
- Many landscape indices are highly correlated (and therefore provide similar or redundant information about the landscape).
- Many studies generate large numbers of metrics and hence many more than the number of degrees of freedom available to facilitate robust statistical analyses.
- Each species responds differently to the same spatial scale of landscape change and human disturbance (e.g. Villard *et al.*, 1999; Davies *et al.*, 2000; Barlow *et al.*, 2007). Hence, no single measure adequately reflects change for all biota.
- Landscape indices may be meaningless when species' perceptions of a landscape are unrelated to the way in which humans characterise and map that landscape.
- Most indices provide an instantaneous, static measure of landscape pattern although temporal dynamics may be important, such as vegetation

succession or the length of isolation of vegetation remnants (Fahrig, 1992; Gascon *et al.*, 1999).

- Values for indices are often scale-dependent, making it difficult to compare results from different landscapes and spatial scales. In addition, the scale of a species' movement is often not well linked to the scale at which landscape indices are generated.

Given these considerations, it is clear that different indices have different strengths and limitations depending on context. The units and the context (e.g. the size of a 'landscape') should be clearly defined. Finally, any system for the application of landscape indices should establish a framework for evaluating predictions and decisions that flow from the application of the indices (Lindenmayer *et al.*, 2002b).

Tests of the peninsula effect

The peninsula effect is an ecological concept derived, in part, from island biogeography theory (Simpson, 1964). A key prediction from this theory is that species richness should decrease from the base to the tip of a peninsula. Several factors may underpin the peninsula effect, including extinction–recolonisation dynamics, variation in habitat suitability along the length of a peninsula and historical factors leading to *in situ* speciation.

The peninsula effect has been tested for many different groups (Wiggins, 1999). These include:

- Birds (Cook, 1969; Raivio, 1988);
- Mammals (Taylor and Regal, 1978; Seib, 1980; Lawlor, 1983);
- Amphibians and reptiles (Kiester, 1971; Busack and Hedges, 1984; Means and Simberloff, 1987);
- Invertebrates (Due and Polis, 1986; Brown, 1987).

Patterns of species richness have not always conformed to predictions from the peninsula effect (Wiggins, 1999), which has motivated some authors to suggest that the concept needs more extensive testing (Means and Simberloff, 1987). In addition, most testing has focused on continental-scale assessments, but if the concept is robust and also generic in nature, then it should manifest at other spatial scales. Current knowledge gaps, together with the spatial scale of some areas of remnant native vegetation that were joined to continuous areas of native forest, provided the stimulus for testing several hypotheses associated with the peninsula effect at Tumut.

Empirical studies of peninsula effects at Tumut

Tubelis *et al.* (2007b) examined evidence for the peninsula effect at Tumut using birds as the target group. They quantified bird species richness in 14 peninsulas: five that were narrow (50–70 m wide), five that were of intermediate width (110–200 m) and four that were wide (300–700 m). To control for other potentially important influences, all peninsulas selected were dominated by the same forest type (Narrow-leaved Peppermint and Ribbon Gum), and the age of the adjacent pine stands was 22–28 years old (Tubelis *et al.*, 2007b).

Bird surveys were then completed within 1 ha plots located 0–200 m, 200–400 m and 400–600 m away from the base of each peninsula (where

(a)

(b)

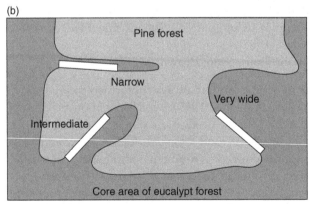

Figure 7.7. (a) Aerial photo of peninsulas at Tumut (photo by David Lindenmayer); (b) Schematic representation of peninsulas surveyed for birds at Tumut (redrawn from Tubelis *et al.*, 2007b).

it connected to continuous areas of native eucalypt forest; see Figure 7.7) (Tubelis *et al.*, 2007b).

Tubelis *et al.* (2007b) found that bird species richness decreased towards the more distal ends of peninsulas. Species richness was higher in wider peninsulas, but the rate of decrease in bird species richness along peninsulas was the same for all width classes. That is, there was an interaction effect for width × location (Figure 7.8).

None of the three general mechanisms listed above that are thought to underpin the development of the peninsula effect at continental levels appeared to be plausible explanations for the bird species richness patterns observed at Tumut. Rather, Tubelis *et al.* (2007b) believe that the factor 'foraging requirements of birds' is the most likely to give rise to lower species richness at the tip of peninsulas.

Lessons from the research at Tumut

The work by Tubelis *et al.* (2007b) demonstrated that the peninsula effect may occur at smaller scales than previously recognised. It is also possible that the value of retained linear strips for biota within production environments (e.g. logged forests; see Recher *et al.*, 1987; Lindenmayer *et al.*, 1993; Zanuncio *et al.*, 1998) might be overestimated if the areas that are surveyed within them are close to contiguous forest. This is because foraging incursions into peninsulas may influence estimates of species richness (Tubelis *et al.*, 2004).

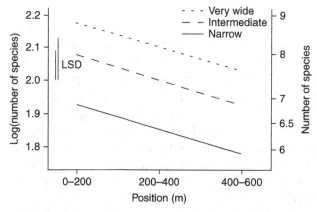

Figure 7.8. Bird species richness along peninsulas of different widths (redrawn from Tubelis *et al.*, 2007b).

Other tests of theory

Data from Tumut have been combined with other large datasets that have been gathered in eastern Australia and used in empirical tests of further ecological concepts, such as those of the resilience of communities and species assemblages (*sensu* Folke *et al.*, 2004) and the landscape texture hypothesis (Holling, 1992). The datasets used in these additional studies include those gathered between 1985 and 2003 over the last 5–23 years in the wet forests of Victoria, the temperate woodlands of southern New South Wales and the coastal heathlands and open forests of coastal Jervis Bay Territory. It is beyond the scope of this book to discuss the results of these additional empirical tests of ecological concepts and theory in detail. However, these studies have led to some interesting new perspectives on ecological theory and concepts. For example, work published by Fischer *et al.* (2007) shows some qualified support for the ecological resilience of bird communities in particular kinds of landscapes such as those dominated by forest cover. There was also support for other ecological theories such as the landscape texture hypothesis (Fischer *et al.*, 2008).

Much work remains to be done on testing ecological theory and concepts. For example, work is currently under way to examine the Tumut and Nanangroe datasets for the presence of assembly rules for birds (Diamond, 1985). Other work on the relationships between animal occurrence and site productivity has recently commenced, as has research on congruence between predictions from the intermediate disturbance hypothesis (Connell, 1978; Shiel and Burslem, 2003) and field data gathered at Tumut.

Summary

Many tests of ecological theory and ecological concepts have been conducted using the datasets gathered at Tumut. As in the case of the work on landscape context effects discussed in Chapter 5, perhaps one of the strengths of the research on ecological theories and concepts was that the work looked beyond the responses of a single group or single level of biological organisation. It spanned different biotic groups (primarily vertebrates) and has focused at different levels of biological organisation – species richness, assemblage composition and individual species responses. These studies have extended the research beyond a simple description of the general findings to better place it in the context of ecological theory and ecological concepts.

Perhaps the most important outcome from the work from an applied perspective has been that it has highlighted the kinds of caveats needed when guiding management to better protect biodiversity. For example, the work on conceptual models has demonstrated not only the need to think deeply about the most appropriate conceptual landscape model to use, but also how the selection of a model may change depending on management objectives. This outcome was not dissimilar to one of the key findings of the work on landscape indices: namely that a single surrogate or small set of surrogates will not reflect the response of all biota to landscape conditions. Similarly, the work on nested subset theory has indicated that simple solutions to conserving species, such as setting aside only the largest remnant patches of forest, may not capture all the taxa of interest; small patches can also make a valuable contribution.

One area of work not examined in this chapter is that loosely associated with concepts of patch occupancy, metapopulation dynamics and population viability. This topic proved to be of major importance in the context of the research at Tumut, and all of the next chapter has been dedicated to discussing it.

8 · Testing PVA models with real data: melding demographic work with population modelling

A substantial body of work at Tumut has involved testing the accuracy of predictions made using generic models for population viability analysis (PVA). The research was among the first to compare field data on patch occupancy by birds, small mammals and arboreal marsupials with predictions of the same measures derived by spatially explicit computer simulation models for metapopulation dynamics.

This chapter contains three sections. The first is a short overview of some background to the testing of PVA models. The second summarises key results for particular animal groups. Findings for each group of species are preceded by a summary of the life history attributes of each taxon to indicate how PVA models were parameterised. The final section of this chapter comprises a general synthesis of modelling outcomes.

Population viability analysis (PVA)

Extinction risk assessment tools such as PVA explore issues associated with the viability of populations (Fieberg and Ellner, 2001). PVA is crucial because the assessment of species extinction risk lies at the heart of conservation biology (Burgman et al., 1993; Fagan et al., 2001). The objective of PVA is to provide insights into how management can influence the probability of extinction (Boyce, 1992; Possingham et al., 2001). It provides a basis for evaluating data and assessing the likelihood that a population will persist (Boyce, 1992; Possingham et al., 2001). More generally, PVA may be seen as a systematic attempt to understand the processes that make a population vulnerable to decline or extinction (Gilpin and Soulé, 1986; Shafer, 1990). In practice, PVA usually refers to building computer-based quantitative models of the likely fate of a population. Probabilities of extinction are estimated by Monte Carlo simulation. The most appropriate model structure depends on the availability of data, the essential features of the ecology of the species and the kinds of questions that managers need to answer (Starfield and Bleloch, 1992; Burgman *et al.*, 1993).

PVA has been widely used around the world and there are hundreds of examples of its use in studies of many species (Beissinger and McCullough, 2002), particularly for rare, declining and threatened taxa (Boyce, 1992; Lindenmayer and Burgman, 2005). While there are several examples where predictions from models and actual dynamics of populations have compared favourably (e.g. Puerto Rican Parrot (*Amazona vittata*), Whooping Crane (*Grus americana*) and Lord Howe Island Woodhen (*Tricholimnas sylvestris*) (see Lacy *et al.*, 1989; Mirande *et al.*, 1991; Brook *et al.*, 2000), in general, PVAs have produced very variable predictions (Coulson *et al.*, 2001; Ellner *et al.*, 2002).

Since Brook *et al.* (2000) assessed the predictive accuracy of PVA for 21 populations (8 bird, 11 mammal, 1 reptile and 1 fish species), discussion about the predictive accuracy of PVAs has increased (e.g. Brook *et al.*, 2000, 2002; Fagan *et al.*, 2001; Fieberg and Ellner, 2001; Beissinger and McCullough, 2002; Reed *et al.*, 2003; O'Grady *et al.*, 2004). Data availability often limits the predictive accuracy of most PVAs (Boyce, 1992; Burgman *et al.*, 1993; Caughley, 1994; Taylor, 1995; Ellner *et al.*, 2002). Even the simplest models require more parameters than are usually available. Even so, PVAs can be valuable in several ways. For example, the use of the models can help organise information for subsequent empirical tests (Walters, 1986), especially if they summarise available data consistently and transparently (Brook *et al.*, 2002). Their use can also help identify knowledge gaps (Burgman *et al.*, 1993), highlight problems for which preemptive action may be beneficial (Tilman *et al.*, 1994) and promote the understanding of complex ecological processes that might otherwise be overlooked in fieldwork (Gilpin and Soulé, 1986; Temple and Cary, 1988; Bender *et al.*, 2003). In addition, through sensitivity analyses, PVA models allow the identification of which model structures and parameters are most important (Possingham *et al.*, 2001).

PVA model testing at Tumut

Most PVA model testing at Tumut has concentrated on the 5,500ha 'pilot study' area in the north-eastern part of the study area at Tumut (see Figure 4.2 in Chapter 4). This subsection of the study area contains 39 remnant eucalypt patches ranging from 0.2 to 40.4ha in size.

Tests of PVA models at Tumut have focused on the accuracy of predictions for arboreal marsupials, birds and small mammals. Extensive fieldwork in the pilot study area, particularly repeated sampling efforts of the 39 patches that it contains, was undertaken to establish patch occupancy and animal abundance for these groups.

Three widely available models were used in simulations of patch occupancy and/or patch-level animal abundance. These were:

- VORTEX (Lacy, 2000);
- ALEX (Possingham and Davies, 1995);
- Hanski's incidence model for metapopulation dynamics (Hanski, 1994b).

The three models are structured quite differently and they simulate population processes in markedly different ways (Lindenmayer *et al.*, 1995b). For example, VORTEX contains detailed sub-programs for modelling inbreeding depression and losses of genetic variability in small populations, whereas ALEX is a more general simulation package focusing on the dynamics of a single (population-limiting) sex – typically females. A detailed description of the models is beyond the scope of this book, and a comprehensive discussion of each one is presented in papers by the respective architects of each model (Hanski, 1994b; Possingham and Davies, 1995; Lacy, 2000).

Information on the life histories of species used to parameterise the models was based on literature values (e.g., fecundity, fledging period) or it came from our field studies in contiguous areas of native forest adjacent to the pine plantation at Tumut (e.g. home range/territory size, population density). The models were **not** parameterised using data from the actual study site. Modellers' predictions can be considered to be like asking someone in 1932 (when the plantation was first beginning to be established) to assess the consequences of the formation of the pine plantation on the distribution and abundance of several species, assuming that they had modern computers, population models and local life history information for each species. This is precisely what the users of PVA models do now to assess the possible consequences of land use development.

Model testing for arboreal marsupials

PVA modelling of arboreal marsupials has involved using VORTEX (Lindenmayer *et al.*, 2000b), ALEX (Lindenmayer *et al.*, 2001b) and Hanski's (1994b) incidence model (Lindenmayer *et al.*, 1999e). Different species were modelled with the different packages – Greater Glider, Mountain Brushtail Possum, Common Brushtail Possum, Common

Ringtail Possum and Yellow-bellied Glider. Not all species were modelled with each package. Of these five species, two sets each of two species were closely related. The Mountain Brushtail Possum and the Common Brushtail Possum are congeneric phalangerid marsupials. The Greater Glider and the Common Ringtail Possum are also closely related and are from the same family (Pseudocheiridae), although one is a marsupial glider and the other is a possum without gliding membranes. Figure 8.1 shows some of the species of arboreal marsupials modelled in this study.

Background biology

All five species of arboreal marsupials examined using PVA models have distributions that include forest and woodland environments that extend well beyond the Tumut region (Dickman and Woodford Ganf, 2007). They vary markedly in life history attributes (e.g. longevity and fecundity), body size, diet and home range (see Chapter 3).

The Mountain Brushtail Possum is a relatively large, non-volant animal with adults weighing up to 4 kg (Viggers and Lindenmayer, 2000; Kerle, 2001). The diet of the species is composed of leaves of several understorey trees and shrubs as well as various types of fungi (Seebeck *et al.*, 1984). Adults tend to live in pairs and occupy a territory of 2–6 ha (How, 1972). The Mountain Brushtail Possum can be long-lived and sedentary and occupy the same area for up to 18 years (Viggers and Lindenmayer, 2000). The species is considered to maintain a monogamous mating system (How, 1972), although this may vary between populations in different parts of the species' distribution (Viggers and Lindenmayer, 2004).

The Common Brushtail Possum is closely related to the Mountain Brushtail Possum (Lindenmayer *et al.*, 2002c), although there are some major differences in the life history strategies of the two species (How, 1981). The species is relatively large and animals may reach 4.5 kg in weight, especially in the southern part of its distribution. The Common Brushtail Possum is omnivorous and its diet includes foliage, flowers, fruit, invertebrates and occasionally birds and their eggs (Kerle, 2001). The home range of the species may vary from less than 1 ha to more than 11 ha depending upon an array of factors such as resource availability and population density (Kerle, 2001). The mating system of the Common Brushtail Possum appears to be polygamous.

The Common Ringtail Possum is a folivore that consumes not only the leaves of eucalypt trees, but also foliage from a range of understorey and shrub species (Pahl, 1984; Wayne, 2005). The species' diet may also

(a)

(b)

Figure 8.1. Images of different species of arboreal marsupials targeted for testing with simulation models for population viability analysis (PVA): (a) Greater Glider; (b) Mountain Brushtail Possum; (c) Common Ringtail Possum (photos by David Lindenmayer); (d)(e) Yellow-bellied Glider (photo by Esther Beaton).

(c)

(d)

(e)

Figure 8.1. (cont.)

include fruits and flowers (How *et al.*, 1984). The home range of the species is typically 1 ha or less (Smith *et al.*, 2003; Munks *et al.*, 2004). The average weight of adults of the Common Ringtail Possum is approximately 750 g. The mating system of the Common Ringtail Possum is unknown, but animals appear to be colonial, with group sizes varying from two to five (Kerle, 2001; Munks *et al.*, 2004).

Adults of the Greater Glider weigh up to 1,300 g and are entirely folivorous, feeding exclusively on the leaves of *Eucalyptus* spp. trees (Kavanagh and Lambert, 1990). The home range of the Greater Glider is about 1 ha (Henry and Craig, 1984; Pope *et al.*, 2004) and the species is usually solitary except during the breeding season, when pairs of animals are often recorded (Henry and Craig, 1984). The mating system of the species may vary between monogamy and polygamy depending on resource availability (Norton, 1988).

The Yellow-bellied Glider is an intermediate-sized animal with adults weighing approximately 650 g (Henry and Craig, 1984). It has highly developed gliding abilities and can volplane (glide) distances of up to 100 m (Goldingay and Kavanagh, 1991). The home range of the Yellow-bellied Glider is 30–60 ha (Craig, 1985; Goldingay and Kavanagh 1993), which is substantially larger than any of the other species of arboreal marsupials which occur in the study area. The Yellow-bellied Glider has a diverse diet which encompasses invertebrates, nectar and sap from eucalypt trees, pollen and honeydew (Goldingay, 1986). The breeding system of the species appears to be variable, changing in response to factors such as resource availability (Lindenmayer, 2002).

Modelling of arboreal marsupials

Model testing using VORTEX involved making predictions using a suite of scenarios including those where:

- The rate of exchange of animals between patches was varied;
- Different models for the migration of animals between habitat patches were invoked;
- Different levels of immigration (or dispersal) from a large neighbouring source area were simulated;
- Variations in habitat quality between remnant patches were incorporated into the model;
- The influence of Radiata Pine stands surrounding the remnant patches was included in model specifications.

Models where the carrying capacity of remnants was assumed to be a simple function of patch area and home range size substantially over-predicted the number of occupied patches and the total abundance of animals (summed across all patches). These results were observed irrespective of the kind of dispersal model and rates of inter-patch movement incorporated into the model. Notably, even these models were substantially more complex than population models used in most other studies to forecast population dynamics in human-modified landscapes. Only when model complexity was increased was better congruence found between predictions from VORTEX and actual values for patch occupancy and overall animal abundance. This was assessed by incorporating the effects of within-patch habitat quality into the model, based on dominant forest type, together with the negative effects of the surrounding pine stands on dispersal mortality (Figure 8.2). Even in those cases

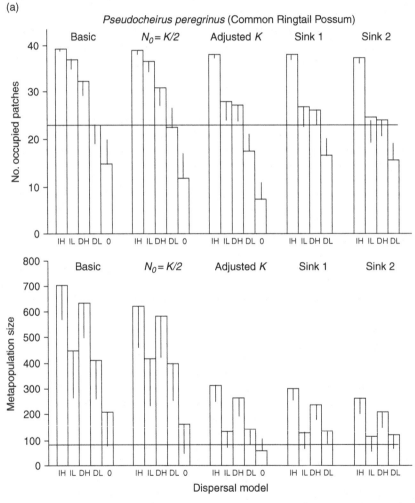

Figure 8.2. Levels of congruence between model predictions from VORTEX and actual values for patch occupancy by arboreal marsupials when the effects of the matrix on dispersal are and are not included in the simulations (redrawn from Lindenmayer *et al.*, 2000b).

where the most complex models were invoked, it was not possible accurately to forecast the number of animals in a given patch. The predicted number of animals was often either well above or well below the observed abundance, possibly as a result of the stochastic nature of extinction and recolonisation events that characterise the dynamics of metapopulations. However, it is also possible that other (unknown) factors not included in the model (e.g. the effects of predation) could have contributed to

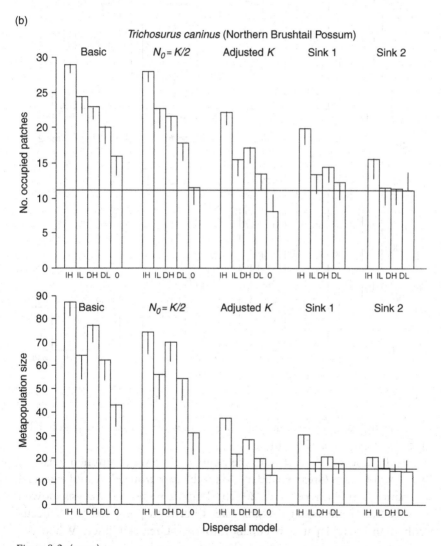

Figure 8.2. (cont.)

(c)

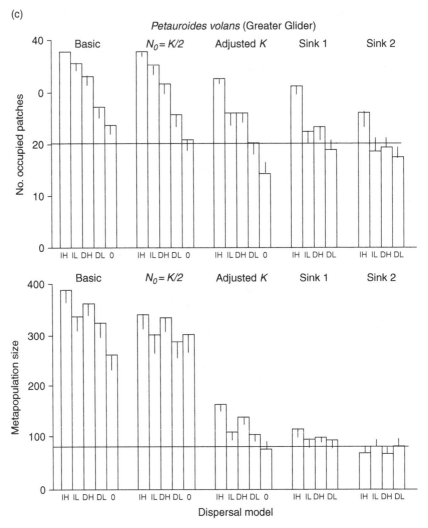

Figure 8.2. (cont.)

the limited predictive ability of the model on a patch-by-patch basis (Lindenmayer *et al.*, 2000b).

In the tests of the ALEX model, congruence was examined between field data for the Greater Glider and predictions from PVA under several scenarios (Lindenmayer *et al.*, 2001b). Logistic regression analysis showed highly significant positive relationships between predicted patch occupancy and actual patch occupancy for the Greater Glider. When the model-derived values for the probability of patch occupancy were high, there was greater congruence between actual patch occupancy and the predicted probability of occupancy. For many patches, probability distribution functions indicated that model predictions for animal abundance in a given patch were not outside those expected by chance. However, for some patches, the model either substantially overpredicted or underpredicted actual abundance. It is possible, therefore, that some important processes, such as inter-patch dispersal, that influence the distribution and abundance of the Greater Glider may not have been adequately modelled (Lindenmayer et al., 2001b).

For tests of Hanski's (1994b) incidence model, patch occupancy was predicted for the Common Brushtail Possum, Mountain Brushtail Possum, Common Ringtail Possum and Greater Glider and then compared against actual data on patch occupancy gathered from extensive field surveys (Lindenmayer et al., 1999e). Data analyses revealed large differences in the results for different species. In the case of the Common Ringtail Possum, there were several models for which it was not possible to derive the parameter estimates required by the model (i.e. relationships for extinction and colonisation as a function of patch size and patch isolation). The ability of the species to persist at low density in the surrounding pine stands matrix may have precluded the calculation of these estimates. There was reasonable congruence between predictions and actual field data for patch occupancy for the Greater Glider and also for the Common Brushtail Possum. Conversely, none of the sub-models for the Mountain Brushtail Possum produced results consistent with the observed data on patch occupancy for this species.

Model testing for small mammals

Background biology

Two species were targeted for detailed analysis – the Bush Rat and the Agile Antechinus (Figure 8.3). The Bush Rat is a native murid rodent with an omnivorous diet of fungi, seeds, fruit, plant tissue and arthropods (Warneke, 1971). The information outlined below is summarised from different field-based studies in south-eastern Australia. Population densities of the Bush Rat can exceed ten animals per hectare. Sex ratios among adults do not differ significantly from parity (Press, 1987). Home ranges of breeding males overlap extensively (Lunney, 1983). Home range sizes vary from 0.1 to 0.4 ha and are typically about 200 m in diameter (Watts and Aslin, 1981; Cunningham et al., 2005).

(a)

(b)

Figure 8.3. Images of (a) the Bush Rat, and (b) the Agile Antechinus, which were targeted for testing with simulation models for population viability analysis (PVA). Photos by Esther Beaton.

More than one litter may be produced in 'good' (wet) years. Litter sizes can vary from one to eight (Warneke, 1971), but the average is about five (Taylor and Horner, 1973). The young are independent from 1–2 months after birth and most do not breed until the following spring (Watts and Aslin, 1981). There is heavy mortality among adults after breeding, and the longevity of most animals is typically less than 15 months. Only a few animals (<5%) survive to breed a second year (Robinson, 1987).

Juveniles remain in the natal territory for 2–3 months before dispersing (Robinson, 1987). No information is available on the distances moved by dispersing individuals. However, animals may move up to 0.5–2km as part of normal nightly movements (Lunney, 1983; Press, 1987), and dispersal distances could be greater than this.

The Agile Antechinus is a small carnivorous marsupial with adults weighing <25g. It feeds on invertebrates, including spiders, beetles and cockroaches (Hall, 1980). Generations of males are discrete because all of them die following an intensive period of mating (Lee and Cockburn, 1985). Many females die after weaning their young and about 20% survive to breed in two successive years. Population densities vary substantially between studies but values of 1–2 animals per hectare were recorded in an investigation completed in an area close to ours (Dickman, 1980). The sex ratio is 1:1 until the males die off. At Tumut, mating occurs in August and all females are pregnant during September (D.B. Lindenmayer et al., unpublished). Litters of between six and ten young are weaned after 3–4 months (Lee and Cockburn, 1985). Agile Antechinus at Tumut typically have six pouch young.

Modelling of small mammals

Model testing for small mammals has been based on ALEX and VORTEX packages. In the case of work using VORTEX, a range of modelling scenarios were completed in which four broad factors were varied (after Lindenmayer and Lacy, 2002). These were:

- Inter-patch variation in habitat quality;
- The pattern of inter-patch dispersal;
- The rate of inter-patch dispersal;
- The population sink effects of the Radiata Pine matrix that surrounded the eucalypt patches.

Model predictions were made for the total number of animals, the distribution of animal density among patches, the total number of occupied patches and the probability of patch occupancy. Predictions were then compared with observed values for these same measures based on extensive field surveys of small mammals in the patch system.

For most simulations using VORTEX for the Bush Rat, the predicted relative density of animals per patch correlated well with the values estimated from field surveys. Predictions of patch occupancy were not significantly different from the actual value for the number of occupied patches in half the models tested. The better models explained 10–16% of the log-likelihood of the probability of patch occupancy. While some of the models gave reasonable forecasts of the number of occupied patches, even in these cases they had only a moderate ability to predict which patches were occupied.

Field surveys revealed that there was no relationship between patch area and population density for the Agile Antechinus – an outcome correctly predicted by only a few models. Five of the 18 scenarios completed for the Agile Antechinus gave predicted numbers of occupied patches not significantly different from the observed number. In each of these five cases, large standard deviations around the mean predicted value meant that uncertainty generated by the simulation model limited the predictive power of the PVA. Some of the models gave reasonable predictions for the number of occupied patches, but those models were unable to predict which patches were actually occupied.

The results of model testing of VORTEX suggested that key processes influencing which specific patches would be occupied were not modelled appropriately. High levels of variability and fecundity drive the population dynamics of the Bush Rat and Agile Antechinus, and Lindenmayer and Lacy (2002) speculated that this may make the patch system at Tumut difficult to model accurately.

Following the work using VORTEX, additional extensive modelling of small mammals was completed using the program ALEX (Ball *et al.*, 2003). The modelling focused on the probability of occupancy of 13 eucalypt forest patches within the Radiata Pine plantation. The modelling was done retrospectively, with simulations commencing at the nominated time point of 1900 – approximately three decades before forest clearing started (in 1932) as part of plantation establishment. Populations of the Bush Rat and the Agile Antechinus were simulated for a time step of 97 years and then data on patch occupancy were compared with field data gathered in 1997 (Ball *et al.*, 2003). Eight different scenarios were run for both species. These were based on differences in inter-patch dispersal, variation in environmental stochasticity and patch size.

Modelling using ALEX produced a range of interesting results (see Ball *et al.*, 2003). In the case of the Agile Antechinus, there was a poor match between predicted and observed patch occupancy for all eight scenarios that were modelled. Patch occupancy was consistently underpredicted by the model. Modelling of the Bush Rat produced marginally better results – one of the eight scenarios accurately predicted the pattern of patch occupancy. This was where the model parameters included low levels of environmental variation, no inter-patch dispersal and large patches of suitable habitat (Ball *et al.*, 2003). Overall, simulations for both species indicated that, while predictions of patch occupancy were poor, in general, those cases where the probability of patch occupancy was predicted to be very high, field data revealed they were actually occupied. The reverse was also true – patches that were unoccupied were often those predicted to have a very low probability of occupancy in the modelling (Ball *et al.*, 2003).

Despite the fact that both the Bush Rat and the Agile Antechinus are among the most studied mammals in Australia, there are attributes of their respective biologies that are at present poorly understood (and which were not included in the VORTEX or the ALEX models) and which could strongly influence patch occupancy. One of these key aspects of biology is the extent of movements such as dispersal. In PVA modelling using parameters for movement based on studies conducted in several other parts of south-eastern Australia (see Lindenmayer and Lacy, 2002; Ball *et al.*, 2003), movement distances of 0.5–2 km were routinely used. This was because these distances can be covered in nightly movements, and hence one of the parameters used in the PVA studies was that dispersal distances could exceed this range. However, in the case of the Bush Rat, subsequent genetic studies (Peakall *et al.*, 2003, Peakall and Lindenmayer, 2006) revealed that these estimates were far too optimistic. The integration of demographic and genetic analyses for this species (see Chapter 9) revealed fine-scale genetic patterns indicative of highly restricted gene flow per generation (>400 m) (Peakall *et al.*, 2003). Both Lindenmayer and Lacy (2002) and Ball *et al.* (2003) noted (although it seemed implausible at the time) that a close fit was found between the simplest model for the Bush Rat, in which there was no dispersal, and observed patch occupancy. Hence, genetic data coupled with PVA modelling suggested that such simple models may indeed be a closer approximation to the real system than was first anticipated. This finding highlights the value of maintaining a range of streams of research (e.g. demographic, genetic and modelling work) in a given area, and illustrates the important synergies and insights that can arise from their integration (Peakall and Lindenmayer, 2006).

Like Peakall *et al.* (2003), Banks *et al.* (2005a, 2005b) also melded demographic and genetic work to derive a better understanding of inter-patch dispersal and patch use, but focused on Tumut populations of Agile Antechinus rather than the Bush Rat. In particular, genetic analyses were used to quantify rates of inter-patch dispersal, especially among males, in the human-modified landscape at Tumut. In addition, alterations to the mating system resulting from landscape modification were also identified by Banks *et al.* (2005a, 2005b). Based on this new information, Banks *et al.* (2005c) then took the work a step further and updated the parameters in VORTEX using an improved understanding of the Tumut landscape, and remodelled patch occupancy by the Agile Antechinus. The new analyses resulted in a considerable improvement in model performance and appeared to represent more accurately metapopulation processes in the Agile Antechinus at Tumut (Banks *et al.*, 2005c). Thus, the work illustrates the value of including an updated understanding of a species' biology in a fragmented ecosystem to improve predictions from PVA models (Banks *et al.*, 2005c).

Model testing for birds

All PVA model testing for birds has been conducted using ALEX. Two major studies have been completed, each involving pairs of relatively closely related species (see Figure 8.4):

- Sacred Kingfisher (*Todiramphus sanctus*) and the Laughing Kookaburra (*Dacelo novaeguineae*) (Lindenmayer *et al.*, 2001c);
- White-throated Treecreeper (*Cormobates leucophaea*) and Red-browed Treecreeper (*Climacteris erythrops*) (McCarthy *et al.*, 2000).

As in the case of work on arboreal marsupials, model testing involved comparing predicted patch occupancy and animal abundance with actual data for these same measures.

Background biology

The Laughing Kookaburra is endemic to mainland eastern Australia. Its diet includes a wide array of invertebrate and vertebrate prey. The majority of nests are in cavities in trees. The species lives in family groups comprising an adult pair in a long-term monogamous relationship, 0–6 auxiliaries or helpers and 0–3 juvenile birds (Higgins, 1999). The adult

(a)

(b)

Figure 8.4. Images of species targeted for testing with simulation models for population viability analysis (PVA): (a) White-throated Treecreeper (photo Graeme Chapman); (b) Red-browed Treecreeper (photo by Julian Robinson); (c) Sacred Kingfisher; (d) Laughing Kookaburra (photos by Esther Beaton).

breeding pair typically produces one clutch per year with a mode of three eggs. A second clutch is occasionally produced, particularly if the first one is lost. Brood reduction, facilitated by siblicide, limits fledging success (Legge and Cockburn, 2000). Little is known about dispersal in the Laughing Kookaburra, although banding records show that animals may move >50 km from the natal territory. However, most marked birds are recovered <10 km from the initial banding site. Mature birds can be long lived, and records of birds exceeding 15 years old are known (Higgins, 1999).

The Sacred Kingfisher occurs throughout large parts of coastal and inland Australia as well as New Guinea, South East Asia, New Zealand and the Pacific Islands. The diet of the Sacred Kingfisher includes a diverse range of prey items including invertebrates and small vertebrates. Nesting usually occurs in tree hollows or termite mounds. The species live in pairs

(c)

(d)

Figure 8.4. (cont.)

and may produce at least two broods annually. Clutch size is typically four and fledging success is about 70% (Higgins, 1999). Populations of the Sacred Kingfisher in southern Australia are migratory and depart in late summer or autumn to move to northern Queensland, New Guinea and Indonesia. Banding records suggest that adult Sacred Kingfishers may return to the same breeding sites in successive years (Higgins, 1999).

The White-throated Treecreeper and the Red-browed Treecreeper both require hollow-bearing trees for nesting and are largely restricted to eucalypt forest. The White-throated Treecreeper appears to prefer rough-barked eucalypts (Lindenmayer *et al.*, 2007d), whereas the Red-browed Treecreeper is predominately associated with smooth-barked eucalypts (Recher *et al.*, 1985). Both species of treecreepers are similar in body size (length = 160 mm and 175 mm, respectively) and have a diet of arthropods taken from the bark of trees (Recher and Holmes, 2000).

There are distinct differences in the breeding biology of the species, with the Red-browed Treecreeper being a co-operative breeder, whereas the White-throated Treecreeper breeds in pairs. The White-throated

Treecreeper appears to have well-developed dispersal capabilities (Doerr and Doerr, 2005; Doerr et al., 2006). In contrast, juveniles of the Red-browed Treecreeper tend to remain close to the natal territory (Noske, 1982, 1986).

Modelling of birds

Modelling of the Sacred Kingfisher and the Laughing Kookaburra produced contrasting results for the two species. For the Sacred Kingfisher, observed patch occupancy was not significantly different from that predicted using the ALEX model. Thus, patches with a high predicted probability of occupancy were often found to be occupied during field surveys. Conversely, for the Laughing Kookaburra, ALEX significantly overpredicted the number of occupied patches, particularly remnants dominated by certain forest types – Ribbon Gum and Narrow-leaved Peppermint. The predictions remained significantly different from observations, even when the habitat quality of these patches was reduced to zero in the model. The rate of dispersal into the system was then changed in the model. This resulted in overall predicted patch occupancy being not significantly different from observations. Nevertheless, the predicted occupancy rates for the different forest types were significantly different from the field observations. The lack of congruence between field data and predictions from the computer models for the Laughing Kookaburra may have arisen because the species was modelled as a metapopulation. However, birds may change their home range movements in response to landscape modification and make use of an array of different patches by moving frequently between them to access spatially separated food and nesting resources (Lindenmayer et al., 2001c).

Modelling of the two species of treecreepers also produced interesting results. The initial models that were employed underestimated the occupancy of the patches, and the models were then modified using the results of the tests in conjunction with further information on the biology of the species. A number of modifications were made to the model to generate results that matched the observations. The best of these modifications made reasonable predictions, although this is not equivalent to a test with independent data because the data were known prior to the modifications. The best-fitting modified models were tested by comparing the observed number of extinction and colonisation events to the predicted number. The models underestimated the observed number of events, although imperfect field survey methods may have contributed to these differences.

The tests of the stochastic models contributed to their development by highlighting the nature of the predictive error. The modified models predicted that the White-throated Treecreeper would be likely to persist over the next 100 years in most of the 40 patches (McCarthy *et al.*, 2000). In contrast, the Red-browed Treecreeper was predicted to become extinct in most patches within approximately 50 years of landscape modification. In the case of the White-throated Treecreeper, predictions were greatly improved if the model included the ability of the species to forage up to 500 m from the remnant eucalypt patches and into the surrounding Radiata Pine stands (McCarthy *et al.*, 2000). Hence, adjustments to the models to account for these effects – increasing their complexity beyond the simple mainland-patch or inter-patch models typically employed in metapopulation modelling – significantly improved predictive ability.

Lessons from Tumut

Table 8.1 summarises the key findings of PVA model testing work at Tumut. The findings demonstrate considerable variation in model predictive ability – both between species for the same model and between models for the same species. Many of the predictions for the Greater Glider were accurate, whereas for two species of small terrestrial mammals (the Bush Rat and Agile Antechinus), the predictions from all the models were generally poor. These are some of the best-studied animals in Australia, and abundant, high-quality life history data are available for them. The results of model testing were attributed to factors such as species-specific relationships between patch size, quality and animal abundance, complex spatial responses to the patch system that could not be readily captured by simple measures of patch size and patch isolation, and the fact that the life histories of the species in altered landscapes were different from those in continuous forests.

All models invariably simplify the systems they attempt to portray (Burgman *et al.*, 1993, 2005). Metapopulation models are no different in this regard (Hanski, 1999) and models such as Hanski's (1994b) incidence function model consistently favour simplicity above model complexity (Ludwig, 1999). Nevertheless, a key conclusion from the work at Tumut has been that it is essential to consider temporal and spatial variation in matrix conditions as part of metapopulation modelling of dynamic managed landscapes. Therefore, it is critical to check model assumptions before models are applied to real landscapes and real conservation problems (Wiens, 1994). The examples of the White-throated Treecreeper and

Table 8.1. *Population modelling outcomes at Tumut. The third column overviews modelling outcomes from the many scenarios and several million simulations completed for each species. Further details of the array of model forecasts tested are given in the accompanying references (modified from Lindenmayer et al., 2003a).*

Group/species	Model[a]	Short summary of outcomes	Reference(s)
Arboreal marsupials			
Greater Glider (*Petauroides volans*)	A	Good congruence with actual patch occupancy and population size	McCarthy *et al.* (2001), Lindenmayer *et al.* (2001b)
Greater Glider	V	Good congruence with actual patch occupancy and population size after adding extra model complexity	Lindenmayer *et al.* (2000b)
Greater Glider	H[b]	Good congruence with actual patch occupancy	Lindenmayer *et al.* (1999e)
Common Ringtail Possum (*Pseudocheirus peregrinus*)	A	Poor predictive ability for patch occupancy and population size, with underprediction even when additional information added to increase model complexity	McCarthy *et al.* (2001)
Common Ringtail Possum	V	Good congruence with actual patch occupancy and population size after adding extra model complexity	Lindenmayer *et al.* (2000b)
Common Ringtail Possum	H[b]	No congruence with actual patch occupancy, impossible to derive parameter estimates required by model	Lindenmayer *et al.* (1999e)
Mountain Brushtail Possum (*Trichosurus caninus*)	A	Poor predictive ability for patch occupancy and population size, with underprediction even when additional information added to increase model complexity	McCarthy *et al.* (2001), Lindenmayer *et al.* (2000b)
Mountain Brushtail Possum	V	Good congruence with actual patch occupancy and population size after adding extra model complexity	Lindenmayer *et al.* (2000b)
Mountain Brushtail Possum	H[b]	No congruence with actual patch occupancy	Lindenmayer *et al.* (1999e)
Common Brushtail Possum (*Trichosurus vulpecula*)	H[b]	Good congruence for patchy occupancy	Lindenmayer *et al.* (1999e)
Yellow-bellied Glider (*Petaurus australis*)	A	Accurately forecast species extinction in patch system	McCarthy *et al.* (2001)

Table 8.1. (cont.)

Group/species	Model[a]	Short summary of outcomes	Reference(s)
Terrestrial mammals			
Bush Rat (*Rattus fuscipes*)	A	Poor predictive ability for actual patch occupancy and population size	Ball *et al.* (2003)
Bush Rat	V	Poor predictive ability for actual patch occupancy and population size, even when additional complexity added to the model	Lindenmayer and Lacy (2002)
Agile Antechinus (*Antechinus agilis*)	A	Poor predictive ability for actual patch occupancy and population size	S. Ball, D. Lindenmayer and H. Possingham, unpublished
Agile Antechinus	V	Poor predictive ability for actual patch occupancy and population size, even when additional complexity added to the model	Lindenmayer and Lacy (2002)
Birds			
White-throated Treecreeper (*Cormobates leucophaea*)	A	Predicted patch occupancy reasonable but underestimated; improved with additional model complexity (e.g. foraging in the matrix)	McCarthy *et al.* (2000)
Red-browed Treecreeper (*Climacteris erythrops*)	A	Predicted patch occupancy reasonable but underestimated; improved with modifications to the model (increased population growth rate)	McCarthy *et al.* (2000)
Laughing Kookaburra (*Dacelo novaeguineae*)	A	Poor congruence for actual patch occupancy with overprediction even after model adjustments (e.g. accounting for habitat suitability)	Lindenmayer *et al.* (2001a)
Sacred Kingfisher (*Halcyon sancta*)	A	Good congruence with patch occupancy	Lindenmayer *et al.* (2001a)

[a] Model codes: A = ALEX, V = VORTEX, H = Hanski's Incidence Model.
[b] Model forecasts patch occupancy and not population size.

Laughing Kookaburra clearly demonstrate that metapopulation modelling and the simulation of populations within patches must take account of the ability of organisms to use, and move across, the surrounding landscape matrix. Incorporating these types of processes into metapopulation modelling will allow this class of models to better account for the hetero-geneity which characterises real landscapes. This will, in turn, increase the value of population models (including metapopulation models) for use in applied natural resource management and biodiversity conservation.

Much of the modelling work completed has compared outcomes for closely related sets of species (e.g. the two species of brushtail possums, the two species of treecreepers). In almost all cases, there were marked differ-ences in findings between each species in a given pair of species. The reasons for the marked differences between species are not clear. However, they indicate that knowledge of the response of one species to disturbance (e.g. landscape modification) may not necessarily provide a useful guide to the possible response of other taxa, including those that are very closely related (Lindenmayer *et al.*, 1999e). This has major implica-tions for the reliability of methods such as indicator species approaches to the conservation of biodiversity (Lindenmayer *et al.*, 2000a). Nevertheless, on the basis of extensive model testing and variations between cases where modelling forecasts were reasonably accurate and where they were not, it was possible to identify the kinds of models and sorts of populations for which predictions are likely to be accurate (see Table 8.2).

Table 8.2. *Contrasts between characteristics of populations whose fates are likely to be more accurately predicted by PVA and those likely to be less accurately predicted (modified from Lindenmayer* et al.*, 2003a).*

More accurate population prediction	Less accurate population prediction
Single population	Metapopulation
Closed population	Open population
Discrete habitat boundaries	Diffuse habitat boundaries
Uniform habitat conditions	Heterogeneous habitat conditions
Constancy of life history attributes across habitat types and landscape conditions	Variation in life history attributes between habitat types and landscape conditions (e.g. fragmented vs. unfragmented landscapes)
Constancy of species interactions across habitat types and landscape conditions	Variation in species interactions between habitat types and landscape conditions
Distance-related dispersal patterns	Habitat-related dispersal patterns
Simple social systems	Complex social systems

Summary

The work at Tumut has indicated that, while PVA models are less than perfect (Lindenmayer et al., 2003a), they nevertheless remain useful tools, particularly in the absence of alternative approaches (Ball et al., 2003). Improvements to PVA models should be motivated by the cycle of testing and continuous upgrading. This is precisely what Banks et al. (2005c) have done with their modelling of Agile Antechinus, where new insights into the biology of the species were used to re-parameterise models for the species at Tumut, and new tests of patch occupancy were preformed – with improved results. Based on these findings, together with those from other kinds of work at Tumut, data on four broad categories of complex spatial processes may be important to add to existing models. These are:

- Relationships between landscape change and key kinds of movement, such as dispersal (Banks et al., 2005a, 2005b);
- Interactions between landscape structure and life history attributes, such as conspecific attraction, as seen for birds at Tumut (Lindenmayer et al., 2004b; see Chapter 6), and altered patterns of space use in habitat fragments (e.g. changed home range size in the Greater Glider; Pope et al., 2004; see Chapter 6);
- Altered patterns of interspecific interactions in modified landscapes, such as elevated predation rates (Murcia, 1995; Ries et al., 2004), as suggested in recent studies of changing landscape context effects at Tumut (see Lindenmayer et al., 2008a);
- Spatially based phenomena such as internal patch population recovery following disturbance (Lindenmayer et al., 2005a).

Tests of PVA models at Tumut have produced some other important insights. First, population modelling has added considerable value to the demographic and genetic data that have been gathered and has pinpointed where additional field and genetic studies may be directed. For example, the lack of congruence between field data and PVA model predictions (Lindenmayer and Lacy, 2002) stimulated the removal experiment that was designed to explore inter-patch dispersal and post-perturbation population recovery (Lindenmayer et al., 2005a, Peakall and Lindenmayer, 2006; see Chapter 9). This further underscores the point made above about the cycle of field data collecting, model testing and model improvement.

Second, the body of work on population modelling was based on a successful and productive collaboration between field ecologists and population modellers – parties with expertise in the biology of the taxa targeted for modelling and the architects of computer programs used in the research, respectively. Burgman *et al.* (1993) and Possingham *et al.* (2001) have argued elsewhere that such a collection of complementary skills and expertise lies at the heart of successful modelling exercises.

9 · Genes in the landscape: integrating genetic and demographic analyses

The potential impacts of landscape change and habitat fragmentation on patterns of genetic variability have been widely discussed in the literature (Manel *et al.*, 2003). However, relatively few large-scale empirical investigations have attempted to link genetic analyses with detailed field-based demographic studies until relatively recently (see Sarre *et al.*, 1995; Saccheri *et al.*, 1998; Keogh *et al.*, 2007; reviewed by Manel *et al.*, 2003). It is even rarer for integrated genetic and demographic analyses to be completed for multiple species in the same landscape (Schmuki *et al.*, 2006a; Kraaijeveld-Smit *et al.*, 2007).

The patch structure and field sampling framework at Tumut provided a useful platform on which to base detailed genetic studies that have been run concurrently with the demographic investigations described in the previous chapters. Genetic studies at Tumut targeted several species:

- The native murid rodent, the Bush Rat (Lindenmayer and Peakall, 2000; Peakall *et al.*, 2003, 2005; Peakall and Lindenmayer, 2006);
- The native marsupial carnivore, the Agile Antechinus (Banks *et al.*, 2005a, 2005b, 2005c);
- The native arboreal marsupial, the Greater Glider (Lindenmayer *et al.*, 1999f; Taylor *et al.*, 2007);
- Two species of native saproxylic beetles (*Adelium calosomoides* and *Apasis puncticeps*) (Schmuki *et al.*, 2006a, 2006b).

This chapter summarises some of the findings of the landscape genetics work on these species. The first section is on the Bush Rat and is the longest because the work included a major removal and population recovery experiment (Lindenmayer *et al.*, 2005a). Integrated genetic and demographic work on the Agile Antechinus is the focus of the second section of this chapter. This is followed by a section on the genetic research on the Greater Glider. An outline of the genetic and demographic work on two species of beetles comprises the fourth and final part of this chapter.

Genetic analyses of Bush Rat populations

Extensive hair-tubing surveys of the Bush Rat at Tumut indicated that the species occupied remnants of eucalypt forest surrounded by extensive stands of Radiata Pine and the probability of occurrence of the species was significantly higher in larger patches (Lindenmayer *et al.*, 1999c; see Chapter 5). The Bush Rat was also a common inhabitant of contiguous eucalypt forest, but was rarely found in stands of Radiata Pine except where windrows of large eucalypt logs had been left following clearing of the original eucalypt forest during plantation establishment.

A series of major field trapping programmes were conducted to gather tissue samples of the Bush Rat and the Agile Antechinus (see below) to facilitate detailed landscape genetic studies of the species. Initially, all 39 patches in the pilot study subsection of the Tumut region (a 10×15 km area in the north-eastern part of the study area – see Chapter 4) were used for trapping, and ear tissue was collected from all individuals of the Bush Rat caught. Subsequent surveys focused more specifically on subsets of eucalypt remnants as well as on areas of contiguous eucalypt forest (Lindenmayer and Peakall, 2000; Peakall *et al.*, 2003).

In the pilot study area that was initially targeted for detailed trapping, significant genetic differentiation was detected among some populations of the Bush Rat. Some neighbouring pairs of remnant populations (less than 0.5 km apart) showed above average genetic differentiation, relative to other more distant populations. Genetic variability was also detected among populations in contiguous native forests. Thus, neither random mating nor isolation by distance could explain the genetic patterns, nor could habitat modification (i.e. plantation establishment) alone account for the findings. This suggested that other processes had contributed to the observed patterns (Lindenmayer and Peakall, 2000). For that initial data-set, no correlations were found between genetic distance and geographical distance in the Bush Rat (Lindenmayer and Peakall, 2000). Instead, populations of the Bush Rat located a considerable distance apart were more closely related if they occurred in patches connected by a drainage line than if they occurred in spatially adjacent populations not connected by a watercourse. The most straightforward explanation for this result was that animals used watercourses for dispersal – there is limited movement among patches not linked via a streamline, even where they are located relatively close to each other (Lindenmayer and Peakall, 2000). This result confirmed the importance of riparian vegetation as a key habitat for small mammals such as the Bush Rat and emphasised the value of its protection, particularly for the maintenance of connectivity between subpopulations.

Patterns of fine-scale spatial genetic structure

The initial work on genetic differentiation in the Bush Rat by Lindenmayer and Peakall (2000) triggered additional genetic studies on movement and dispersal, and hence gene flow, in the species (Peakall *et al.*, 2003, 2005). Much of this was based on intensive trapping of transects established in two creek gully sites within large areas of contiguous native forest and in two eucalypt remnants surrounded by stands of Radiata Pine plantation (Peakall *et al.*, 2003). As in earlier work on the Bush Rat, animals were captured during repeated trapping events using metal box (Elliott) traps, and a small piece of ear tissue was used for subsequent genetic analyses. A substantial part of these genetic analyses involved the use of microsatellite markers combined with multilocus autocorrelation methods to investigate fine-scale (>1 km) patterns of spatial distribution of genetic structure that resulted from dispersal movements in the Bush Rat (Peakall *et al.*, 2003).

The detailed additional genetic studies of the Bush Rat revealed a number of highly unexpected findings. In particular, there was strong evidence of fine-scale genetic structure for all study sites and for both sexes and all age classes of animals (i.e. sub-adults and adults). Thus, there was no evidence of random genotypes in the populations that were studied. Rather, proximate animals were more alike than distant ones and were typically distributed in clusters which were approximately 200 m or less in diameter, interspersed with gaps of lower density (Peakall *et al.*, 2003). These patterns of spatial genetic structure were consistent with limited dispersal and hence restricted gene flow in the species. The genetic evidence for restricted per-generation dispersal was further supported by mark–recapture analysis of animals in which the distance moved by genetically tagged animals had a mean value of only 35 m (±4.5 m) (Peakall *et al.*, 2005). These corroborated findings of limited gene flow were all the more unexpected because trapping studies by other workers had indicated that some animals may make 1–2 km 'sallies' in a single night (Lunney, 1983). Importantly, such patterns of restricted gene flow were **not** simply an outcome of landscape modification (i.e. plantation establishment) because they were observed not only in the eucalypt remnants but also at sites surveyed within contiguous forest (Peakall *et al.*, 2003).

The work by Peakall *et al.* (2003, 2005) was notable for the reason that it was the first time that patterns of positive fine-scale spatial genetic structure had been reported for a small mammal. The research was also focused at a scale that was different from that used in most population genetic studies at that time (typically tens to hundreds of kilometres) and, as a result, contributed new insights into the ecology of the Bush Rat, particularly the dispersal biology of the species. This, in turn, proved to have a substantial impact on interpreting the findings of the major experimental removal study described in the following section.

A population removal and recovery experiment on the Bush Rat

Data on patch occupancy in the Bush Rat showed high levels of variability in animal abundance between patches (see Chapter 5). These observations, in particular the absence of animals from some of the remnants, stimulated an experimental removal study. This involved a controlled and replicated field experiment linked to a genetic study to examine the rate and mechanism of population recovery after severe experimental population reduction (Lindenmayer *et al.*, 2005a). Specifically, the following questions were asked:

- Do Bush Rat populations recover following severe depletion? If so, over what period?
- Is the rate of population recovery in a given patch related to patch size or the distance to other habitat fragments?
- Bearing in mind the combined insights provided by demographic and genetic data, what is the most likely mechanism of population recovery?

Experimental design

Of the 192 eucalypt remnants within the Radiata Pine plantation at Tumut (see Chapter 4), 24 were chosen for the removal experiment. A further six eucalypt remnants in an adjacent agricultural landscape were also included in the study as release sites for translocated animals (see below).

Eucalypt remnants were selected that were more than 3km apart to avoid problems with homing behaviour (Fox and McKay, 1981). Eucalypt patches surrounded by pine stands containing large windrowed logs were excluded because such windrows support limited numbers of small mammals. Therefore, a series of remnant patches were selected for which it was assumed (with reasonable confidence) that few, if any, small mammals lived in the surrounding pines.

The main factors considered in the manipulative experiment were perturbation treatment, patch size and patch isolation. The four levels of perturbation treatment were:

- **Removal patches**, where the aim was to clear patches of all animals through intense and thorough trapping over the whole site. Trapping continued until no more animals were captured and patches were considered to be devoid of trappable animals. Captured animals were then translocated to other eucalypt patches in the system.
- **Removal and restocking patches**, where the same comprehensive trapping as in removal-only patches took place but the patches were 'restocked' with new animals brought in from other eucalypt remnants.
- **Control patches**, where trapping took place on a grid only, because these remnants were often slightly larger than other patch types. Captured animals were marked and then released at the exact point of capture.
- **Release patches** where, following trapping, animals were released into remnant patches located in an adjacent agricultural area at least 3 km from their source population.

Two levels of isolation were considered: 'near' (<200 m to other euca- lypt patches) and 'far' (>500 m to other eucalypt patches). Given logistic constraints, the patch sizes for the study were constrained to between 1.3 and 2.8 ha and so were not considered as a treatment factor, but rather as a covariate. As far as possible, an attempt was made to replicate, at least once, each of the treatment combinations. The assignment of patches to treatment combinations was randomised.

At the beginning (March 2000) and end (March 2002) of the study, an attempt was made to capture all small mammals in each of the eucalypt patches, using traps placed approximately 10 m apart. Trapping con- tinued in each patch until no new animals were captured for at least two successive nights. A minimum of five nights of trapping was under- taken in each patch, but sometimes the number of nights of trapping exceeded 20.

At the same time as complete patch trapping was undertaken, a sub-area of each patch was established within each patch, comprising a permanent grid of 25 traps set 10 m apart. These grids were usually located in the centre of each patch. The permanent grids were trapped four times: as part of complete patch trapping work in March 2000 and March 2002, and as part of grid-only trapping in August 2000 and March 2001.

Thus, grid sampling was conducted at 0, 6, 13 and 24 months into the study period.

Genetic analyses

As part of the capture protocol, a small piece of ear tissue was collected from each captured animal for genetic analyses. Genetic analysis was subsequently applied at 11 hypervariable microsatellite loci (Peakall *et al.*, 2003). Genetic tagging, which is the identification of individuals by genetic markers, was also conducted (Peakall *et al.*, 2005). This enabled unequivocal identification of individual Bush Rats and ensured that recaptures could be distinguished from animals caught only once. This approach was crucial to ensuring that the demographic data were robust and reliable.

Demographic data and post-perturbation population recovery

Large inter-patch differences in population size were recorded at the commencement of the project, with some patches supporting very large numbers of animals (more than 50) despite being very small in size. Analyses of grid data provided evidence that a significant knockdown in population size occurred. After 6 months, the average number of animals on removal sites was 58% lower than on the 'control' sites. At initial trapping for the same set of grids, removal sites had 30% fewer animals than the controls. Furthermore, there was evidence of a significant relationship between counts after 6 months and initial counts for the control sites. This was not the case for the removal sites. However, count data gathered after 24 months for all sites showed that the relationship between final and initial census counts was almost directly proportional and did not appear to be strongly affected by perturbation treatment. That is, overall, the recovery process was complete after 24 months – patch populations that were initially small subsequently recovered to a similarly small size, and those that were initially large were also large 24 months later. Analysis of counts obtained by grid trapping at 0, 6, 13 and 24 months showed that populations of the Bush Rat were beginning to recover after 6 months and that there was no evidence of a difference in overall temporal patterns in log counts between treatments.

Irrespective of whether the study patches were close to other eucalypt patches or were isolated, there was no significant change in the numbers of animals before and after treatment at 24 months. There was some weak evidence for greater recovery of small populations in larger patches, but this trend was not statistically significant.

Genetic analyses of population knockdown

The above demographic evidence for a knockdown effect in the removal experiment was supported by genetic data. There was a significant genetic change between the pre- and post-treatment removal populations. Conversely, no significant genetic change was detected between trapping sessions without perturbation (controls). The mean genetic differentiation between the pre- and post-treatment 'perturbed removal' populations was higher than the mean differentiation between trapping sessions without perturbation, but was not significantly different. The mean genetic differentiation between the pre-treatment populations and their nearest neighbour was significantly higher than for both the average perturbed and average unperturbed pair-wise comparisons. Therefore, while significant genetic change occurred as a consequence of the perturbation, the magnitude of genetic change was only marginally greater than what would occur naturally between trapping sessions in the absence of perturbation. Notably, the magnitude of genetic changes between pre- and post-treatment was significantly less than the average genetic differentiation between neighbouring populations. That is, the post-treatment populations were genetically more similar to their respective pre-treatment populations than to their neighbouring populations.

Mechanisms underpinning rapid post-removal population recovery

Two plausible mechanisms for the rapid recovery of populations of the Bush Rat were:

- **Immigration**. Immigration into patches where animals had previously been removed has been observed in some other studies of mammals (e.g. Middleton and Merriam, 1981; Stenseth and Lidicker, 1992a).
- **Natural population growth** from offspring produced by residual animals that escaped capture and removal could have helped return animal abundance to pre-treatment levels. Although an intensive trapping effort was employed to remove all animals from patches, it is still possible that some remained untrapped. Indeed, one of the potential problems in removal experiments is that some residents remain in the 'cleared' area and when trapped later are incorrectly classified as recolonists (reviewed by Stenseth and Lidicker, 1992b). The Bush Rat is a classic r-selected strategist (*sensu* MacArthur and Wilson, 1967) along the r–K continuum (Pianka, 1970) and the species is characterised by

high levels of fecundity (i.e. they can produce large litters) – life history attributes that would encourage rapid population recovery.

These two mechanisms produce clear predictions that aid discrimination between them. If immigration is the predominant mechanism, a strong relationship is predicted between the rate of recovery and distance from source populations (Hanski, 1994a). At the same time, relative to recovery by residual animals, the recovery of populations exclusively by immigration would be expected to be relatively slow. Also, under immigration alone, a negative relationship between rate of recovery and patch size would be expected, while the opposite is likely with recovery by residual animals since the number of residual animals left behind is likely to increase with increasing patch size, merely as a result of the increasing logistic difficulty of trapping entire larger populations.

Empirical evidence of full recovery within 24 months, combined with a lack of evidence for any isolation effects, were both consistent with recovery by residual animals, rather than recovery by immigration. It thus seems likely that the high fecundity of the Bush Rat enabled populations to recover – even if only very small numbers of animals remained following perturbation. Despite this, it was difficult from demographic data alone to discriminate between the alternative mechanisms of population recovery. However, multiple lines of genetic evidence were consistent with the hypothesis of recovery from residual animals. First, assignment tests (Paetkau et al., 2004) across the four largest populations in the removal treatment revealed that 99% of all post-treatment animals could be genetically assigned to the pre-treatment population. Thus, residual animals left a discernable genetic 'signature' in the recovered populations and they (together with their offspring), rather than colonists from neighbouring populations, formed the major component of the recovery. Second, while patterns of genetic change detected after the experimental perturbation confirmed a knockdown effect, the magnitude of the genetic change was intermediate relative to the average genetic differentiation among neighbouring populations. Such a genetic pattern was consistent with recovery from within the population. Third, other genetic work on Bush Rat populations at Tumut (see above) revealed fine-scale genetic patterns indicative of highly restricted gene flow per generation (Peakall et al., 2003), leading to predictions of limited recolonisation following local population depletion or extinction (Peakall and Lindenmayer, 2006).

Lessons from the work

The removal experiment generated many new insights that had not been anticipated at the start of the study. The rate and completeness of recovery were striking and highly unexpected, particularly given the extent of the initial population knockdown. The limited dispersal distances of the Bush Rat were also highly unexpected, particularly given the results from other workers showing that individual animals may move more than 1–2km in a single night (Lunney, 1983; Cunningham et al., 2005).

The results of the removal experiment have some important implications for the persistence of biota in landscapes modified by humans. First, while the role of immigration in population reinforcement and recovery is well known (e.g. Brown and Kodric-Brown, 1977), the importance of residual animals (as identified in this study) is less well studied and is rarely included in models of population dynamics. Second, residual animals can be considered to be 'biological legacies' and they clearly play an important role in the recovery of disturbed populations and ecosystems (e.g. Turner et al., 2003; Lindenmayer et al., 2008c). Biological legacies are defined by Franklin et al. (2000) as organisms, organically derived structures and organically produced patterns that survive from the pre-disturbance system. In the removal experiment, biological legacies (in the form of a small number of residual animals) remained in some patches and significantly influenced population recovery patterns (Peakall and Lindenmayer, 2006).

These unexpected results for limited gene flow in the Bush Rat suggest an interesting ecological conundrum. That is, while populations of the Bush Rat appear to be relatively resilient to disturbance and can recover quickly, localised extinction may in fact be difficult to reverse. Thus, this common and widespread species may be maladapted to the kinds of landscape modification that lead to disrupted habitat connectivity for the Bush Rat (see also Tilman et al., 1994). It may be that common and widespread species such as this have never needed to evolve well-developed dispersal capabilities because they were resident in landscapes with high levels of natural connectedness (*sensu* Noss, 1991).

Genetic analyses of Agile Antechinus populations

Extensive trapping and hair-tubing data for the Agile Antechinus at Tumut indicated that the species was almost never captured within sites dominated by stands of Radiata Pine. However, the species was found in approximately 50% of the 192 eucalypt patches that occur in the study area (Chapter 5). In addition, isolation effects were identified for patch occupancy (Chapter 5). Based on these observations, a series of major integrated ecological and genetic studies of the species was instigated by Banks *et al.* (2005a, 2005b, 2005c). The work involved trapping animals and collecting tissue samples for genetic analyses in a wide range of patches of native forest within the plantation and sites in contiguous native forest. These surveys were complemented by an additional investigation in which trapping, radio-tracking and genetic analyses of animals was completed in four small eucalypt patches, four large eucalypt patches and four sites in large areas of contiguous eucalypt forest.

The work explored a wide range of effects of landscape modification on the Agile Antechinus including those on dispersal, social organisation, kin structure and the risk of inbreeding. The remainder of this section summarises a small subset of the many valuable findings from the research.

Initial animal capture work by Banks *et al.* (2005c) revealed that 70% of eucalypt patches that were trapped were occupied by the Agile Antechinus. This contrasted with results from contiguous areas of native forest in which all (100%) of trapped sites supported the species (Banks *et al.*, 2005c). Statistical analyses indicated that patches least likely to be occupied were those that were a long distance away from contiguous areas of native forest. Genetic analyses based on gene diversity and allelic richness at microsatellite and mitochondrial markers found no evidence for the complete isolation of patch populations (Banks *et al.*, 2005c). Thus, populations in the eucalypt patches were not simply isolated remnants of the population that existed prior to human modification of the landscape to establish the plantation.

Other studies by Banks *et al.* (2005a, 2005b, 2005c) focused more strongly on the movement, behaviour and breeding ecology of the Agile Antechinus, using trapping, radio-tracking and genetic techniques. A relatively detailed description of the results can be found in Chapter 6 because they are highly relevant to how patch use can be altered in highly modified landscapes. Some of the most important points are listed below for the sake of completeness.

- Dispersal was male-biased in all environments, i.e. eucalypt patches and areas of continuous native forest. By contrast, females remained in the natal territory, irrespective of landscape context (except pine stands, where the species does not occur).

- Landscape modification resulted in eucalypt patches supporting fewer males than in areas of contiguous native forest, an outcome of impaired male dispersal across pine stands.
- Isolation effects on dispersal were pronounced for distances exceeding 750 m between eucalypt patches and areas of contiguous native forest, but limited for distances of 250 m or less.
- Gene flow was restricted where animals had to disperse across stands of Radiata Pine. However, strips of native vegetation along gully lines facilitated dispersal.

Banks *et al.* (2005a) found that the landscape context-induced restrictions in dispersal underpinned some important changes in the breeding biology of the Agile Antechinus.

- Levels of relatedness among potential mates were elevated in eucalypt patches.
- Sharing of nests in trees with hollows by opposite-sex relatives was limited in large eucalypt remnants but not in small eucalypt remnants.
- Reductions in the size of populations in the eucalypt patches, particularly for males, reduced the incidence of multiple paternity of litters.

These changes mean that the effects of landscape modification and habitat fragmentation in species such as the Agile Antechinus are not limited simply to extinction–colonisation dynamics, but can also include altered breeding behaviour, with potential knock-on impacts on recruitment, inbreeding depression and eventually population viability (Banks *et al.*, 2005a, 2005b).

Genetic analyses of Greater Glider populations

Integrated genetic and demographic studies of the Greater Glider were an exciting and unique component of the work at Tumut. In particular, the availability of genetic samples extracted from animals alive during a phase of plantation establishment in the 1960s was a rare opportunity to quantify landscape modification effects. On this basis, the work on the Greater Glider aimed to produce a better understanding of connectivity and dispersal and how it may have been influenced by human landscape modification. Two bodies of information indicated that the Greater Glider would be a useful case study taxon.

First, before the commencement of the study of the Greater Glider, it was anticipated that the species would be highly sensitive to disturbances such as plantation establishment because:

- It cannot persist in stands of Radiata Pine, as these areas do not provide food (the species is almost totally dependent on eucalypt leaves) or shelter (the species is an obligate hollow-user and cavity trees are absent in exotic pine stands in Australia).
- Long-distance gliding movements would be impeded in dense pine plantations (Suckling, 1982).
- The species is sensitive to extreme temperatures (Rübsamen et al., 1984) and is dependent for daytime shelter within tree-hollows, which are not present in Radiata Pine stands (Gibbons and Lindenmayer, 2002).
- The body mass of the Greater Glider is at the lower limit for an arboreal folivore and adult animals have little margin to meet either their minimum nitrogen or energy requirements (Foley and Hume, 1987; Foley et al., 1990), so that access to eucalypt leaves during long-distance movement is essential.

Second, extensive surveys of 86 eucalypt remnants at Tumut (see Chapters 4 and 5) revealed that approximately 20% of patches still supported the species (Lindenmayer et al., 1999a). This was a significantly lower site occupancy rate than the 38% of the 40 matched 'control' sites in large contiguous areas of eucalypt forest adjacent to the Radiata Pine plantation (Chapter 5). These findings strongly suggested that the Greater Glider may have been lost from some eucalypt patches since they first became surrounded by pine stands, and that recolonisation does not occur readily.

Despite apparent evidence that plantation establishment has impeded glider movements, other observations are contradictory. First, there were so few gliders (less than ten) in some patches that they would be unlikely to be viable in the long term unless they experienced immigration. Second, in a pilot study of five remnant patches, 11 gliders that had initially been marked disappeared from the patches in 12 months, and 15 new gliders appeared (Pope, 2003; Cunningham et al., 2004b), suggesting that there may be considerable turnover, either from other more distant patches or from the surrounding contiguous eucalypt forest. Thus, a key to understanding extinction risks in the Greater Glider in this system was to elucidate whether patch populations have survived in isolation since plantation establishment, or whether they are supplemented by immigrants from other patches and/or the eucalypt forest surrounding the plantation.

Hypothesis testing prior to genetic analyses

The existence of Greater Glider specimens collected prior to plantation establishment in the 1960s, the persistence of animals in remnant eucalypt patches and the occurrence of the species in surrounding 'control' areas of contiguous native eucalypt forest provided a unique opportunity to examine a range of questions associated with changes in genetic variability associated with temporal and spatial landscape change. Thus, prior to the onset of detailed genetic analyses, a number of hypotheses were constructed about the possible impacts of landscape modification on the Greater Glider. These hypotheses were based on general conservation biology and conservation genetics principles and were published prior to the commencement of genetic studies of the species (Lindenmayer et al., 1999f). Specifically, two key questions were addressed (after Taylor et al., 2007):

- Do individuals of the Greater Glider in isolated eucalypt patches display genetic signatures consistent with being small, isolated remnants?
- Does geographical context affect rates of immigration into patches?

Collection of historical samples

Numerous specimens of the Greater Glider were collected in the 1960s in the study area as large areas of native eucalypt forest were cleared to establish stands of Radiata Pine (Tyndale-Biscoe and Smith, 1969a, 1969b). The precise locations where the specimens were collected were recorded before the specimens were subsequently lodged with the Australian Museum, National Museum of Victoria, South Australian Museum and the National Wildlife Collection (CSIRO Wildlife and Ecology). Samples of hair, skin and teeth were taken from these historical specimens held at various museums, and DNA was extracted from them.

Field captures of the Greater Glider

The nocturnal Greater Glider is active from <7 to 30+ metres above the ground. It is difficult to lure into traps, given its widely available diet of leaves (Tyndale-Biscoe and Smith, 1969a, 1969b; Kavanagh and Lambert, 1990). Individuals were captured on the ground after using a rifle to prune branches on which animals perched (Kehl and Borsboom, 1984). Captured animals were sedated and blood samples collected (Viggers and Lindenmayer, 2001). Animals were released at the exact point of capture once they had fully recovered from the effects of the sedative (Viggers and Lindenmayer, 2001).

Field sampling of current populations of the Greater Glider

Field survey work described in Chapter 5 showed that of a total of 86 remnant eucalypt patches spotlighted for the Greater Glider, 18 contained animals. Six of these were too close to human habitation to discharge a rifle and two were so heavily infested with Blackberry that it was impossible to make effective captures of animals. All gliders in the remaining ten patches (N=2–18 gliders in each) of eucalypt forest ranging in size from 1.6 to 30 ha were captured (Figure 9.1). Repeated spotlight surveys on a series of successive nights by four highly experienced observers ensured that all gliders in these patches were detected

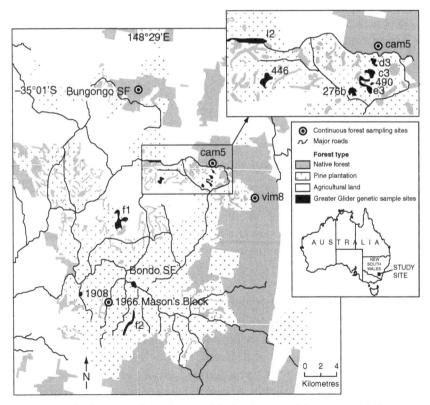

Figure 9.1. The location of sites from which the Greater Glider was sampled for genetic analyses (shown in black). Redrawn from Taylor *et al.* (2007).

and subsequently caught. Less intensive surveying was also conducted in a 124 ha patch (see f1 in Figure 9.1), from which seven individuals were caught and sampled.

Five of the eucalypt patches in the north of the plantation (c3, d3, e3, 490 and 276b), were close to contiguous eucalypt forest and had been isolated 15–18 years previously. These patches were separated from one another by distances of less than 1 km, and were the site of the study by Pope et al. (2004) (see Chapter 6). The other three sites in the northern part of the plantation had also been isolated 15 years before, but were more distant from other sampled areas (f1, i2, 446) (Figure 9.1). Three southern sites had been isolated 33–34 years before and were between 3 and 7 km apart (see Bondo, 1908 and f2 in Figure 9.1). The Greater Glider was also captured at three sites located within large continuous areas of eucalypt forests that surrounded the pine plantation.

Genetic analyses

To determine the contribution of patch connectivity to the persistence of the Greater Glider within eucalypt remnants, 12 microsatellite markers were used to genotype individuals from 11 remnants, 3 contemporary nearby contiguous native eucalypt forest sites and a sample collected during native vegetation clearance prior to plantation establishment in the 1960s.

Genetic analyses showed that samples from eucalypt remnants retained substantially more genetic diversity than expected if animals in patches were totally isolated. This indicated that some inter-patch movement had occurred. Six putative patch immigrants – two from sampled sites 1 and 7 km distant, and four from unresolved or unsampled localities – were identified via genetic parentage and population assignment analyses. Although animals in remnant habitat patches typically showed some reduction in genetic diversity, the loss was less extreme than might be expected for very small aggregations isolated for the 4–8 generations since clearing of the surrounding native forest and its conversion to an exotic Radiata Pine plantation, where these animals cannot survive. Under drift/mutation equilibrium (Crow and Kimura, 1970), the effective population sizes required to maintain heterozygosity at the observed levels, in the absence of immigration, range from 6 to 37 adults – much larger than the current census sizes for animals in the relevant patches (Taylor et al., 2007).

Although no group of gliders appeared to have been completely isolated since the surrounding stands of Radiata Pine had been established, genetic data suggested that some locations had experienced less immigration resulting in gene flow than others. Patch populations displayed varying levels of admixture – with the oldest and most geographically isolated ones showing the least admixture, suggesting that they had experienced relatively little immigration. Evidence of at least some immigration into patches may explain why the Greater Glider has persisted contrary to expectation in heavily fragmented habitat.

In summary, the use of high-resolution genetic data and analyses was a valuable approach to developing critical new insights into patterns of connectivity in the Greater Glider. The study revealed that although the Radiata Pine plantation surrounding the eucalypt remnants did not support populations of the species, some individuals nevertheless dispersed through these stands of exotic trees. Thus, inter-patch connectivity ensured that some patches of remnant eucalypt forest continued to be occupied by the species, including those that were relatively isolated from extensive areas of continuous eucalypt forest (Taylor *et al.*, 2007).

Genetic analyses of saproxylic beetle populations

One of the more unusual studies at Tumut over the past decade has been a combined ecological and landscape genetics study of two species of saproxylic or dead-wood-inhabiting beetles (*Adelium calosomoides* and *Apasis puncticeps*; see Figure 9.2) (Schmuki *et al.*, 2006a, 2006b). Both are flightless, wood-boring tenebrionids that are closely associated with rotting logs.

Schmuki *et al.* (2006a) collected over 2,300 beetles from 142 rotting logs located in 52 sites at Tumut that were divided roughly evenly between areas of contiguous eucalypt forest and patches of remnant native forest surrounded by pines. They then focused analyses on patterns of spatial genetic structure, allelic richness and the density of beetles per m^3 of wood.

The analyses revealed many interesting findings (Schmuki *et al.*, 2006a), but only a small subset of these are discussed below. First, there was evidence of male-biased dispersal in *Adelium calosomoides* but not *Apasis puncticeps*. Second, *A. calosomoides* and *A. puncticeps* showed evidence of restricted gene flow in patches of remnant native forest compared with sites sampled within contiguous native forest. Such effects were apparent in a number of measures, including spatial genetic structuring, allelic richness and patterns of isolation by geographical distance. Thus, stands of Radiata Pine impaired gene flow in both species of beetles. For example, allelic richness in *A. puncticeps* was negatively associated with distance from contiguous native forest – it decreased by 0.113 km ($\pm$0.035 km) for

(a)

(b)

Figure 9.2. (a) *Adelium calosomoides;* (b) *Apasis puncticeps* (photos by Christina Schmuki).

every kilometre away from contiguous forest (Schmuki *et al.*, 2006a). Third, the paucity of logs and leaf litter in pine stands appeared to be the primary reason for reduced mobility. Finally, in addition to pine stands, landscape characteristics such as roads and streamlines appeared to be barriers to the dispersal of *A. calosomoides* and *A. puncticeps* and impeded gene flow (Schmuki *et al.*, 2006a).

Summary

Traditional ecological methods such as radio-tracking and colour-banding can be limited in their ability to describe dispersal patterns at landscape scales and, in the absence of long-term monitoring, they are uninformative with respect to population history (Dieckmann *et al.*, 1999). Thus, genetic methods such as microsatellite genotyping can help to provide an improved ecological understanding of responses to landscape modification (e.g. Berry *et al.*, 2005; Keogh *et al.*, 2007). Indeed, the work summarised in this chapter highlights the value of working across disciplines – the melding of genetic and demographic data helped to tackle fundamental questions about post-disturbance population recovery, animal dispersal in modified landscapes and how the life histories of species change as a consequence of human landscape modification.

Many highly unexpected findings were brought to light as a result of landscape genetic analyses at Tumut. First, one of the most widespread species, the Bush Rat, showed a strong spatial genetic structure that was indicative of limited dispersal. This led to the surprising conclusion that this common and widespread species, which is capable of rapid recovery following disturbance (Lindenmayer *et al.*, 2005a), may in fact be extinction-prone in highly modified landscapes (Peakall and Lindenmayer, 2006). Second, at the start of work at Tumut, the Greater Glider was predicted to be one of the most dispersal-limited species in the region because of its inability to live in pine stands. However, genetic analyses uncovered strong evidence of more movement than was initially anticipated. Third, landscape modification triggered marked and not readily predictable changes in the life history of the Agile Antechinus, particularly in the social organisation and breeding behaviour of the species (Banks *et al.*, 2005a, 2005b, 2005c).

The new insights derived from genetic analyses have helped explain some of the results from work described in earlier chapters of this book. For example, limited movement in the Bush Rat and the Agile Antechinus is likely to be a key reason why predictions of patch occupancy using PVA models proved to be so inaccurate (see Chapter 8). Similarly, improved ecological understanding drawn from genetic analyses of the Greater Glider has helped to explain why some small and remote patches of eucalypt forest were occupied by the species (Chapter 5).

10 · *Refining and extending the research programme: additional studies at Tumut (and nearby) that build on the Fragmentation Study*

The Tumut Fragmentation Study has spanned basic research on the demographics of vertebrate populations, testing of ecological theory, population modelling and genetics – all of which have implications for conservation biology as well as for forest and plantation management. However, as the study evolved, it became apparent that there existed some limitations of these initial studies. This is hardly surprising – perfect studies don't exist. Even famous large-scale experimental studies such as the Biological Dynamics of Forest Fragments Project in Brazil (Laurance *et al.*, 1997; Gascon *et al.*, 1999) have statistical and ecological limitations.

A key deficiency at Tumut was that the lack of animal and plant occupancy data of eucalypt patches prior to major landscape modification (i.e. plantation establishment) weakens the inferences that can be made about biotic responses to change (Margules, 1992). Such a limitation is common in the vast majority of studies of landscape change and habitat fragmentation worldwide (Lindenmayer and Fischer, 2006).

A second deficiency was the rapidity with which the stands of Radiata Pine surrounding the patches of native eucalypt forest were modified. The dynamic nature of the age and condition of pine stands significantly influenced the biota which inhabited the eucalypt patches (Tubelis *et al.*, 2004). Hence, the results derived from work completed, for example, in 1996 or 2000 may not reflect responses at a subsequent time when pine stands surrounding a given eucalypt remnant are clear-felled. Indeed, one of most substantial challenges in establishing the original study encompassing 166 sites at Tumut was finding the 40 pine sites that were to remain unaltered for 2–3 years (see Chapter 4).

This chapter summarises two additional studies at Tumut, and another major linked study in the nearby Nanangroe area, that attempted to overcome of the deficiencies of earlier studies at Tumut. All three are more experimental than the observational cross-sectional study described in detail in Chapters 4 and 5 of this book.

The Edge Experiment

The plantation estate at Tumut is highly dynamic, with stands subject to repeated thinning operations every ~10 years, as well as a clear-felling operation at the end of a 25–30 year rotation. Clear-felled stands are then replanted with young Radiata Pine trees and the growing cycle recommences. Remnant patches of native eucalypt forest are exempt from logging but their context changes as the surrounding pine stands are clear-felled. In 2000, a new study was instigated to explore the impacts of a changing (clear-felled) landscape matrix on remnant patches of native eucalypt forest. The work was assigned the name the 'Edge Experiment' and results from it have begun to emerge in the past year. The design and preliminary results of the study are briefly outlined below.

Experimental design

A simple replicated before–after control–impact (BACI) experiment was implemented to quantify the response of vertebrates to a change in the matrix surrounding native forest patches. From the pool of 192 eucalypt remnants surrounded by Radiata Pine stands in the Buccleuch State Forest at Tumut, 24 eucalypt remnants were selected for detailed study. The 24 remnants were grouped as 12 matched pairs. Each pair consisted of one patch where the surrounding stands of exotic Radiata Pine were clear-felled (termed 'treatment patches'; Figure 10.1) and one matched eucalypt patch (termed 'control eucalypt patches') where surrounding pine stands remained unlogged for at least a further 5–7 years. Patch size, climate, forest type and geology data were used to cross-match the 24 sites in the study. This ensured that the range of environmental and other conditions were matched across the treatment and control patches.

Birds and arboreal marsupials were surveyed for a year beforehand (2001) and then several years repeatedly after pine stands were cut around the eucalypt remnants (2002–2006). Control and treatment patches were surveyed at the same time to avoid confounding treatment and time.

(a)

(b)

Figure 10.1. Examples of: (a) a 'treatment eucalypt patch' where the surrounding pine stand has been clear-felled (photo by Darius Tubelis); (b) a 'treatment eucalypt patch' where the surrounding pine stand has been clear-felled on one side. (Photo by David Lindenmayer.)

Observers were randomly allocated to sites to avoid observer × treatment confounding. The entire experiment continued for 5 years (four surveys) but had to be terminated in 2006 when a large part of the Buccleuch State Forest was burned in a wildfire and the fire-damaged pine stands surrounding control patches were subjected to salvage logging.

Results of the Edge Experiment to date

Analyses of the data gathered for birds revealed highly significant differences between sites whose surroundings were fully cut and the control sites. Species richness was reduced by approximately 4–9 species in patches where the surrounding pine stands had been clear-felled. Highly significant relationships were found between a negative response to clear-felling of the surrounding pine stands and the nest type. Birds with cup and dome nests were those negatively affected by cutting of the pine stands surrounding eucalypt treatment patches. On the other hand, birds that nested in hollows or in suspended purses remained relatively unaffected by clear-felling of the surrounding stands of plantation pine. Birds with cup nests may be susceptible to altered microclimatic conditions that typically arise when adjacent areas are clear-felled areas. An additional or alternative explanation might be that cup-nesting birds are more susceptible to nest predation when the surrounding pine stands are removed than species which nest in tree hollows or suspended purses.

In contrast to the findings for birds, the abundance of two of the three most common species of arboreal marsupials in the Edge Experiment (the Common and Mountain Brushtail Possums) showed negative responses to partial cutting; a significant negative effect which persisted when surrounding pine stands on both sides of a patch were clear-felled. The relatively limited mobility of arboreal marsupials may explain why negative effects of changes in the surrounding pine matrix become apparent when only part of it is clear-felled.

The Edge Experiment has shown that a number of native vertebrate taxa will persist in the remnant areas of eucalypt forest, even when significant changes take place in surrounding plantation stands (see Chapter 11). This emphasises the importance for wildlife conservation of maintaining areas of native vegetation within the boundaries of industrial plantations. In the particular case of birds, more species will be maintained within eucalypt patches if the scheduling of logging is amended so that not all of the surrounding pine plantation is clear-felled in a single operation.

The Nest Predation Study

An increased level of predation on bird nests is a phenomenon observed frequently in the forests of the Northern Hemisphere (Murcia, 1995; Kremsater and Bunnell, 1999; Ries *et al.*, 2004). At Tumut, a study of nest predation was conducted in different types of edge environments. Rates of predation of quail eggs deposited in artificial nests (Figure 10.2) that were set out at varying distances away from roads were measured. The nests were located at sites in different types of forest: extensive stands of Radiata Pine, large areas of contiguous *Eucalyptus* forest, and patches of remnant native forest surrounded by stands of Radiata Pine.

Four broad categories of sites were established (see Figure 10.3), each characterised by differences in the type of surrounding forest:

Figure 10.2. Artificial nest used in the nest predation experiment at Tumut. (Photo by David Lindenmayer.)

- Remnant patches of eucalypt forest surrounded by stands of Radiata Pine on three sides and on a fourth side separated by a gravel road from large contiguous areas of eucalypt forest;
- Remnant patches of eucalypt forest surrounded by stands of Radiata Pine on three sides and on a fourth side separated by a gravel road from areas of Radiata Pine plantation;
- Large contiguous areas of eucalypt forest bisected by a gravel road;
- Extensive areas of Radiata Pine bisected by a gravel road.

Eight replicates of each of the four broad categories of sites were established, giving 32 sites in total. The sites were located at least 2 km apart to reduce the potential for spatial dependence in the results. Artificial nests were placed at distances of 0, 50, 100, 150 and 200 m along an unmarked 'transect' set each side of the gravel road which characterised each site. Thus, at any given site, a 400 m-long transect with ten nests was established. Standard markers used in forest survey, such as flagging tape and spray paint,

Roadway context – Remnant and contiguous eucalypt forest

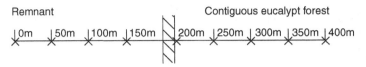

Roadway context – Remnant and Radiata Pine

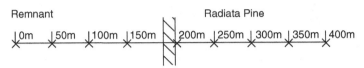

Roadway context – Radiata Pine

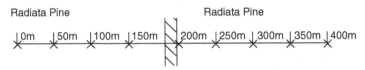

Roadway context – Contiguous eucalypt forest

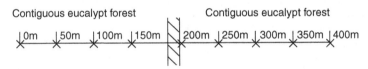

Figure 10.3. Experimental design and site types in the nest predation experiment at Tumut.

were not used in the study in order to avoid the possibility of providing obvious visual cues for potential predators. Instead, the position of each nest was carefully mapped to enable it to be quickly relocated.

The design structure of the experiment included the following measures: category of site, type of forest within a site, distance from the road and plant species in which an artificial nest was attached. However, statistical analyses showed that none of the design or measures variables in the study were statistically significant.

The key outcome of this study was that the rate of egg predation from artificial nests did not vary in response to contrasting types of forest edge environments or distance from roads into the forest; it was essentially random. Based on the results of these analyses, the predicted probability of predation on an artificial nest was 35% (95% CI = 28–44%). The findings contrasted with those of other workers (see review by Forman *et al.*, 2002). For example, variation in nest predation with edge depth has been recorded in many Northern Hemisphere studies (Paton, 1994). It is possible that populations of some potential predators are ubiquitous throughout the Tumut study region and they exert similar levels of predation pressure in the different types of areas that were examined.

Although the investigation was a designed field study with site replication in which many nests were deployed, there are limitations in the work completed (and indeed many others where artificial nests have been used; see Major and Kendal, 1996). For example, the study was confined to a limited period of time (2 weeks) and employed artificial nests that were not attended by adult birds – two factors that can affect the results of such experiments (Major and Kendal, 1996). In addition, data were lacking on predation levels on natural nests so that it was not possible to contrast them with the observed rates for artificial ones.

The Nanangroe Natural Experiment

As outlined at the start of this chapter, a major deficiency of the work on landscape context effects at Tumut was that it was a cross-sectional or 'snapshot' study (*sensu* Diamond, 1986). That is, it involved surveys during a single time period or set of time periods within sites located in three broad landscape context classes (see Chapters 4 and 5). Recognition of this deficiency led to the instigation in 1997 of a large-scale longitudinal study called the 'Nanangroe Natural Experiment' in a nearby area targeted for plantation conversion.

The work at Nanangroe is examining the impacts of plantation expansion on biota inhabiting woodland patches when they are surrounded by stands of Radiata Pine. The Nanangroe experiment is a long-term

---

Box 10.1. The strength of longitudinal studies

Often interest is in the direct study of temporal changes and relationships that occur as a result of intervention. In these cases, long-term (longitudinal) studies are necessary as they can distinguish changes over time from differences between sites in initial population levels (e.g. Davies *et al.*, 2000, 2001). That is, cohort effects can be separated from age effects, which were confounded in the work on landscape context effects at Tumut. Furthermore, as each site becomes its own control, a longitudinal study may provide convincing evidence of relationships among key variables, which may not be discerned from data obtained in cross-sectional studies.

(longitudinal) study with the defining feature that repeated observations have been (and will continue to be) taken on individual sites, enabling the direct study of change.

The Nanangroe Natural Experiment focuses on a grazed woodland landscape near Jugiong in south-eastern New South Wales, southern Australia, approximately 15 km north-east of the boundary of the Tumut study area. It contrasts strongly with other 'fragmentation experiments' where extensive clearing of trees occurred to isolate habitat fragments (e.g. Margules, 1992; Laurance et al., 1997; Bierregaard et al., 2001). Rather, it is examining a landscape that is already largely cleared of the original vegetation cover and which now supports only remnants of the native eucalypt woodland. The landscape surrounding these woodland fragments is undergoing a rapid transition from one that is largely cleared and dedicated to grazing domestic livestock to one dominated by extensive plantations of Radiata Pine trees. The work at Nanangroe is quantifying the changes that occur in vertebrate fauna inhabiting woodland remnants when the surrounding landscape undergoes extensive change. The groups targeted for study are birds, terrestrial mammals, arboreal marsupials, reptiles and frogs.

Experimental design

In 1997, a set of foundation sites for the study at Nanangroe was established. All 70 patches of remnant native vegetation that occurred on land broadly designated for subsequent pine plantation establishment were surveyed and their characteristics recorded. In 1998, prior to the commencement of landscape transformation to a pine-dominated system, a random selection of 52 of the 70 woodland remnants was made from strata defined by vegetation class and woodland patch area. These woodland remnants were exempt from clearing as the plantation was established and are hereafter referred to as the 'woodland treatments'. Four patch size classes were considered within these 'woodland treatments', and the site selection procedures yielded 13 remnants in the 0.5–0.9 ha size class, 20 remnants in the 1.0–2.4 ha class, 17 remnants in the 2.5–4.9 ha class 2 and 2 remnants in the 5.0–10.0 ha class. Few large remnants were available for selection because of the extent of previous clearing for livestock grazing.

Three broad woodland vegetation classes within the 'woodland treatments' were recognised:

- Red Box and Red Stringybark (co-dominant) with Apple Box, Long-leaf Box (*Eucalyptus goniocalyx*) and Broad-leaved Peppermint (*E. dives*) (19 sites);

- Yellow Box, White Box, Red Stringybark (co-dominant) and Blakely's Red Gum (23 sites);
- Mountain Swamp Gum and other kinds of vegetation (e.g. River Oak, *Allocasuarina cunninghamiana*) (10 sites).

Surveys of vertebrates commenced before stands of Radiata Pine were planted, and the sites were then surveyed repeatedly after landscape treatments had been implemented. Clearing to plant Radiata Pine in the areas surrounding the woodland remnants involved the felling and burning of isolated paddock trees and other shrubby vegetation (Fig. 10.4). Two age cohorts of recently planted pine stands were defined. These were trees established in 1998 (Cohort 1) and trees established in 2000 (Cohort 2). Of the 52 'woodland treatments', 11 had one or two open boundaries with adjacent agricultural areas. Radiata Pine stands completely surrounded the remaining 41 woodland remnants.

In addition to the 52 'woodland treatments', a set of 'natural control' sites was established. These included:

- Ten sites in the newly planted stands of Radiata Pine trees (hereafter referred to as 'pine controls');
- Fifty-six woodland remnants on semi-cleared private grazing properties surrounded by areas of scattered paddock trees and located adjacent to the new plantation estate (hereafter referred to as 'woodland controls');
- Ten permanent sites in grazing paddocks that surrounded the 56 woodland remnants on the grazing properties (hereafter referred to as 'paddock controls').

Domestic livestock grazing continued in all woodland remnants in the study and the maturing Radiata Pine stands, thus preventing potential confounding between treatments and grazing effects. Table 10.1 summarises the 'treatment' design and number of sites in each category.

In summary, six key features characterised the design of the Nanangroe study:

- Quantification of animal abundance prior to landscape treatments being applied. Each site then became its own control, making the investigation more powerful than cross-sectional studies for studying temporal change.
- The use of a randomised and replicated patch selection procedure to reduce the potential for bias and to average over random factors.
- The establishment of a set of 'natural external control' sites to quantify natural year-to-year fluctuations in animal occupancy of sites. This also provided additional contrasts and hence helped to better quantify patterns of change.

(a)

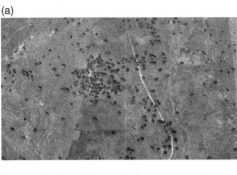

(b)

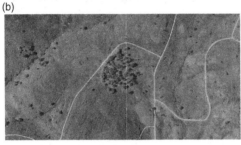

(c)

Figure 10.4. Landscape changes on the Nanangroe property between 1997 and 2005 showing landscape context changes around site NAN28 approximately in the centre of each image. (Photos courtesy of Forests NSW, Tumut Office.)

- The repeated measurement of biophysical attributes at the site level to link changes in these to changes in biota.
- Repeated field surveys over a prolonged period to better quantify species response trajectories to landscape transformation.
- The large spatial scale, which made it appropriate for mobile groups such as birds (Wiens, 1999), which can be difficult to investigate with small-scale studies (Debinksi and Holt, 2000).

Table 10.1. *Number of sites cross-classified by site type, cohort, context, number of edges and year of planting, highlighting the factorial structure of the design. Counts of sites for the first (1998) and eighth (2005) years of the study are presented to show minor losses or additions of sites as plantation establishment proceeded.*

Site type	Surrounding vegetation	Cohort	No. of edges	Year 1998	2005
Landscape context change					
Remnant	Pine	1	1–2	3	3
Remnant	Pine	1	3–4	16	15
Remnant	Pine	2	1–2	8	5
Remnant	Pine	2	3–4	25	29
'Controls'					
Remnant	Paddock	–	–	56	55
Pine	–	–	–	10	10
Paddock	–	–	–	10	10

Results of the Nanangroe Natural Experiment to date

Most macropod, arboreal marsupial and reptile species showed evidence of approximate linear increases in occupancy rates within the 'woodland treatments' over the 9-year period 1997–2006 (see Figure 10.4 for an illustration of the landscape changes occurring over that period). Similar trends occurred in the 'woodland controls', suggesting that the species had increased throughout the region, rather than displaying year × treatment interaction effects.

For birds, complex and varied significant treatment effects were recorded. Bird responses were not clearly associated with life history attributes such as body size, diet or foraging guild. Birds were cross-classified according to their type of temporal response to landscape transformation. Groups included:

- Species that either increased or decreased across the entire region – a broad-scale regional change;
- Species that decreased within the 'woodland treatments' but, at the same time, increased in the 'woodland controls'– possibly a displacement effect;
- Additional (primarily forest) bird species recruited to the 'pine controls' – an added-environment effect;

- Bird species which either decreased or increased within the 'woodland treatments' in comparison with the 'woodland controls' – a longitudinal landscape context effect.

Attributes of the 'woodland treatments' significantly associated with changes in bird occupancy were:

- The age of surrounding pine stands;
- The number of boundaries a woodland patch had with surrounding pines;
- The size of the woodland patches;
- The dominant vegetation type of the woodland patches;
- Temporal changes in the vegetation structure and plant species composition of the ground and understorey layers of woodland treatments.

Although the results from the work at Nanangroe were consistent with other studies that highlight the importance of the retention of native vegetation remnants within plantation-dominated regions, several novel and unanticipated effects were identified. First, region-wide changes were observed for many species of vertebrates. Thus, the impacts of landscape transformation possibly extended well beyond the broad areas where the treatments were applied. Second, no functional group relationships were found that could be linked with responses to landscape context effects.

Collectively, the findings to date suggest that several key processes are likely to be drivers of change at a range of spatial scales. At the largest scale, overall region-wide changes were identified in animal occupancy rates that were positive for some taxa and negative for others. At a patch scale, an array of species were influenced by changes in the landscape context of woodland remnants when the surrounding landscape was transformed to pine stands. At the smallest spatial scale, localised changes in habitat attributes within woodland patches surrounded by maturing pines had a significant influence on some species. Recognition of these processes appears to be important for proactive conservation management in landscapes transformed by the widespread phenomenon of broad-acre plantation expansion.

Summary

Several investigations have been completed that are directly or indirectly related to the main cross-sectional study (see Chapters 4 and 5) that has formed the backbone of the research programme at Tumut. Two key ones are the Edge Experiment and the Nanangroe Natural Experiment, which were, in part, established to overcome some of the deficiencies in the cross-sectional study.

Both the Edge Experiment and the Nanangroe Natural Experiment have produced interesting new insights that were not readily predicted from earlier work at Tumut. For example, key results for particular mammal and bird species at Tumut (Lindenmayer *et al.*, 1999a, 2002a) could not be readily transferred to the Nanangroe ecosystem. The Common Ringtail Possum was abundant at Tumut, and at the start of the Nanangroe work it was anticipated that the species would increase in the woodland remnants as areas of surrounding pine matured. However, the reverse trend occurred and the species declined across the entire study area. Similarly, the Sulphur-crested Cockatoo and Sacred Kingfisher were common birds in the remnant forest patch system at Tumut (Lindenmayer *et al.*, 1996, 2002a) but exhibited study-area-wide declines at Nanangroe.

In the case of the Edge Experiment, predictions of bird species responses based on earlier studies of the avifauna in the Tumut region proved to be inaccurate. Landscape context analyses (see Chapter 5) identified particular sets of bird species most likely to occur in stands of Radiata Pine. These included birds belonging to particular foraging guilds such as sweepers and ground carnivores (*sensu* Mac Nally, 1994) and those nesting closer to the ground (Lindenmayer *et al.*, 2002a). At the start of the Edge Experiment, it was anticipated that bird species with these life history attributes might be those disadvantaged by the clear-felling of stands of Radiata Pine. However, a surprising outcome was that neither foraging guild nor nesting height were life history attributes common to those bird taxa which declined significantly following clear-felling. All of the bird species which declined were taxa which, as previous studies in the Tumut region had shown, made extensive use of pine stands for foraging (Cunningham *et al.*, 1999; Lindenmayer *et al.*, 2002a; Tubelis *et al.*, 2004), although they rarely breed there. Eucalypt patches may have been too small to provide sufficient food to support these species and they disap-peared once adjacent areas in which they foraged were removed.

The unexpected outcomes from both the Nanangroe Natural Experiment and the Edge Experiment suggest there is a need for humility about how much is really known about biotic responses to landscape modification (see also Mac Nally *et al.*, 2000). They also underline the fact that findings from one place or time might not be readily transferable to other places or other times in the same broad area or region.

11 · Recommendations for plantation managers: implications for biodiversity and conservation in plantations

The preceding chapters have highlighted the extensive and diverse body of scientific research that has taken place at Tumut over the past 12 years. The work was originally intended as a largely theoretical landscape ecology and conservation biology study, primarily focused on quantifying landscape context effects on biota (see Chapters 4 and 5). However, as the research programme has progressed, it has become increasingly clear that the findings from the research at Tumut (as well as those from the nearby Nanangroe area; Lindenmayer *et al.*, 2008b; see also Chapter 10) have implications for biodiversity conservation in plantation landscapes. Some of these lessons might apply not only elsewhere in Australia (Lindenmayer and Hobbs, 2004; Salt *et al.*, 2004), but also in the extensive areas of plantation elsewhere in the world. This chapter briefly outlines a set of recommendations arising from the Tumut research that aim to improve conservation management practices within plantations.

The significance of plantation expansion as a form of landscape change

Around the world, large-scale landscape change resulting from human land use (UNEP, 1999) is a major driver of altered ecosystem processes (McIntyre and Hobbs, 1999; Millennium Ecosystem Assessment, 2005) and biodiversity loss (Sala *et al.*, 2000; Lindenmayer and Fischer, 2006). Plantation expansion is a significant form of landscape transformation worldwide (Jackson *et al.*, 2005) and it is an increasingly common land use in places such as South America (Estades and Temple, 1999), North America (Haskell *et al.*, 2006), Europe (Shakesby *et al.*, 1996; Martínez-Sánchez *et al.*, 1999), Asia (Cubbage *et al.*, 1996), Japan (Yamaura *et al.*,

2006), Australia (Burns *et al.*, 1999; Salt *et al.*, 2004), New Zealand (Clout and Gaze, 1984; Allen *et al.*, 1995; Dyck, 2000) and Africa (Wethered and Lawes, 2003).

In 1996, the global area of plantations exceeded 130 million ha (Cubbage *et al.*, 1996), and in 2001 it was more than 187 million ha (Food and Agriculture Organization of the United Nations, 2001, 2007). Indeed, there is a worldwide trend towards a greater reliance on wood sourced from plantations (Food and Agriculture Organization of the United Nations, 2001, 2007), with an increasing emphasis on 'industrialised' plantation forestry in the Southern Hemisphere (Franklin, 2003; Barlow *et al.*, 2007). Increasingly, plantations are also being established to help offset carbon emissions through carbon sequestration (Jackson and Schlesinger, 2004).

A better understanding of the responses of biota to landscape transformation, including large-scale plantation establishment and expansion, is pivotal to the development of effective natural resource management strategies (Peterken and Ratcliffe, 1995; Moore and Allen, 1999; Lindenmayer and Hobbs, 2004). The work at Tumut has generated some key findings that illustrate the critical importance of well-informed plantation design based not on the development of large-scale monocultures of introduced trees, but rather on the creation of heterogeneous landscape mosaics (*sensu* Bennett *et al.*, 2006), where stands of planted trees are interspersed with patches of remnant native vegetation. Such considerations are among those which form a key part of integrating conservation and production in plantation environments.

Softwood plantation management and the conservation of remnant native vegetation

Landscapes dominated by softwood plantations have wood and paper production as their primary aim. However, some changes in land-use protocols have the potential to contribute significantly to the conservation value of these landscapes. The following protocols and guidelines are based on the findings of the various studies that have been completed at Tumut. Many of these recommendations apply to existing plantations as well as new ones such as those (like Nanangroe) being established on semi-cleared land that was formerly used for grazing and which support patches of remnant native vegetation.

Remnant native vegetation within plantations

The importance for biodiversity of patches of remnant native vegetation has been illustrated by several studies at Tumut. Their value was quantified for several groups including arboreal marsupials, small mammals, birds, reptiles, amphibians, invertebrates and plants (Chapter 5). Larger patches of remnant native vegetation, particularly those greater than 3 ha, supported more species of native animals than smaller patches. Patch size effects were identified for many groups, including arboreal marsupials (Lindenmayer *et al.*, 1999a, 2000), birds (Lindenmayer *et al.*, 2002a), reptiles (Fischer *et al.*, 2005) and some (but not all) kinds of invertebrates (Smith, 2006) (see Chapter 5).

Although larger patches supported more species than small ones for many groups, work at Tumut and Nanangroe has demonstrated that remnant patches of native vegetation as small as 0.5 ha have considerable value for biodiversity (Parris and Lindenmayer, 2004; Lindenmayer *et al.*, 2008b). Moreover, traditional kinds of increasing patch size and species richness relationships were notably absent for some groups such as frogs (Parris and Lindenmayer, 2004) and bryophytes (Pharo *et al.*, 2004). Furthermore, analyses of patterns of nestedness at Tumut suggest that small patches can often support species not found in large patches (Fischer and Lindenmayer, 2005b; see Box 7.1, Chapter 7), and a range of patches and patch types may be required to represent the complete assemblage of species found in the area. Based on these collective findings, a general recommendation to land managers is that remnant patches of native vegetation that are 0.5 ha or larger should not be cleared as part of softwood plantation establishment.

Work at Tumut and at Nanangroe has demonstrated that the composition of groups such as birds and arboreal marsupials can vary significantly between different vegetation types that are dominated by different species of eucalypt trees (Lindenmayer *et al.*, 1999a, 2002a; Chapter 5). Given this, strategies for native vegetation retention within plantations need to ensure that a range of different vegetation types are conserved. This is consistent with recommendations based on studies of reptiles at Tumut (Fischer *et al.*, 2005) to maintain landscape heterogeneity encompassing patches of remnant native vegetation of different sizes, shapes and vegetation types (dominated by different species of native trees).

Remnant patches at Tumut that were close (<250 m) to large continuous areas of native vegetation were more likely to be occupied by some taxa (e.g. small mammals and invertebrates) at Tumut than those that were more isolated (>750m away) (Banks *et al.*, 2005a, 2005b, 2005c; Schmuki *et al.*, 2006a). Therefore, patches of remnant eucalypt forest close to large areas of continuous eucalypt forest are those most critical for retention during plantation establishment. However, even isolated patches can have significant conservation value for many groups such as birds (Lindenmayer *et al.*, 2002a) and they should not be cleared simply because, from a human perspective, they are perceived to be isolated.

Any clearing of native vegetation as part of plantation establishment will have negative impacts on biodiversity. However, it is a practical reality that plantation establishment will entail some clearing. For example, plantation establishment in many landscapes in south-eastern Australia will take place on former grazing properties and require clearing of scattered paddock trees which can otherwise have a range of ecological roles (Manning *et al.*, 2006). A general recommendation to land managers is that an offset policy should be adopted in these circumstances (Gibbons and Lindenmayer, 2007) in which the removal of these trees is countered by native vegetation restoration efforts elsewhere in the plantation estate, particularly along gully lines (see below). However, clearing existing areas of native vegetation and establishing new plantings elsewhere will typically have short- to medium-term negative effects on many elements of biota. This is because recently established tree plantings lack many of the key stand structural attributes such as large trees with hollows and large fallen logs that are critical habitat attributes for many species (Cunningham *et al.*, 2007).

Extent of remnant vegetation within plantation boundaries

A landscape comprising a mosaic of remnant native vegetation and soft-wood stands will have significantly higher biodiversity value than a Radiata Pine monoculture (Friend, 1982a, 1982b; Lindenmayer *et al.*, 2002a). Simulation modelling (McCarthy and Lindenmayer, 1999) of the landscape at Tumut suggested that extensive new developments of soft-wood plantations (e.g. >1,000ha) should aim to contain at least 30% of remnant or re-established native vegetation. Restoration efforts may be needed to reach such levels of native vegetation cover within a plantation. However, empirical data from both Tumut and Nanangroe suggest that smaller amounts of remnant native vegetation than this can have a sub-stantial positive effect on the suite of native species that persist in an area (Lindenmayer *et al.*, 2002a, 2008b).

Spatial and temporal patterns of patch–matrix contrast

The work at Tumut, and more recently at Nanangroe (see Lindenmayer *et al.*, 2008b; Chapter 10), has demonstrated in many different ways that the landscape context of retained patches of native vegetation matters. Chapter 5 highlighted the significant landscape context effects quantified for virtually all the broad groups of plants and animals examined at Tumut. In the particular case of the work at Tumut, some species of conservation concern appeared to benefit from plantation establishment. An example was the Rufous Whistler – a species thought to be declining elsewhere, particularly in woodland environments in south-eastern Australia (Barrett *et al.*, 2003). This result for the Rufous Whistler was mirrored in the longitudinal experiment at Nanangroe (Lindenmayer *et al.*, 2008b).

The work at Nanangroe has more directly demonstrated landscape context effects and, in particular, how the transformation of landscapes significantly alters the presence and abundance of species and the composition of assemblages inhabiting woodland remnants surrounded by maturing pine stands. In that study, the general findings for birds suggested a shift from woodland-dominated bird assemblages to assemblages that are a unique combination of forest and woodland birds (Lindenmayer *et al.*, 2008b; Chapter 10). In contrast, the composition of mammal and reptile assemblages has remained largely unchanged, although most species appear to have increased in abundance in the woodland remnants where the landscape context has changed through the establishment of pine stands in the surrounding areas.

It is difficult to make general recommendations to plantation managers about the changes in biotic assemblages within patches of retained vegetation associated with changes in landscape context. They do need, however, to be aware that although some species will be gained (primarily forest ones), others may be lost, including some of conservation concern.

While the Nanangroe study has demonstrated that the establishment of pine stands in an area can have a significant impact on woodland patch occupancy (Chapter 10), the results of the 'Edge Experiment' indicate that the complete removal of the pine matrix surrounding patches of remnant native vegetation can also have significant impacts on some species of birds and mammals (Chapter 10). A recommendation from that work is that, where possible, the cutting schedule of clear-felled plantation stands adjacent to remnants should be staggered to ensure that, at any one time, eucalypt remnants are adjacent to some areas of advanced regrowth Radiata Pine.

Management of gullies

Many studies have shown that riparian areas and gully lines have higher levels of species richness and greater abundances of particular species than other parts of landscapes (Recher *et al.*, 1980; Soderquist and Mac Nally, 2000). Although equivalent results were not identified in the work at Tumut, riparian areas and gully lines were nevertheless found to be important habitats for a wide range of species from many groups, including birds, mammals and frogs. The integration of detailed demographic and genetic data obtained from studies of small mammals at Tumut indicated that gully lines were likely to be dispersal routes for the Bush Rat (Lindenmayer and Peakall, 2000) and the Agile Antechinus (Banks *et al.*, 2005a) (see Chapter 9).

Given the importance of riparian vegetation as habitat and for wildlife dispersal, a general recommendation for management is that such areas should be exempt from clearing during the establishment of new plantations. In existing plantations where riparian vegetation was cleared in the past, gully lines should be targeted for restoration after the completion of the final clear-fell operation.

Management of aquatic areas

Maintenance of the integrity of aquatic environments is a key guiding principle for the conservation of forest biodiversity (Lindenmayer *et al.*, 2006, 2007a). This holds both for native forest and for plantation landscapes. Work at Tumut indicated that attributes of water bodies such as reeds, rushes, sedges and emerging and fringing vegetation had a significant effect on frog species richness (Parris and Lindenmayer, 2004). Degradation or loss of these attributes of water bodies can lead to the loss of habitat for many individual species of amphibians. Thus, streams, wetlands and other aquatic areas need to be exempt from conversion to plantations. They also need to be buffered from the effects of plantation establishment in the adjacent landscape (Parris and Lindenmayer, 2004).

Hygiene of plantation machinery

Large populations of invasive plant species characterise both the Tumut and Nanangroe plantation estates (Chapter 5) as well as exotic softwood plantations elsewhere in south-eastern Australia (Burdon and Chilvers, 1994) and around the world (Richardson *et al.*, 1994). Blackberry thickets are common in pine stands and neighbouring eucalypt remnants. Radiata

Pine wildlings have become established in many areas of remnant native vegetation surrounded by plantation pine (Lindenmayer and McCarthy, 2001; Chapter 5). Three key recommendations for invasive plant species arise from the work conducted at Tumut to date:

- Hygiene protocols are needed for logging and other machinery to stop the spread of Blackberry from the existing plantation estate (where it is already established) to other areas targeted for the development of new plantation forests.
- Management effort is required to control presently minor outbreaks of Blackberry in recently established parts of the plantation estate (e.g. those planted in the past 5 years) to prevent such outbreaks from becoming major infestations. This is because control efforts are most likely to be effective at the early stage of investigation (McNeeley et al., 2003).
- Efforts should be made to further develop and use reproductively sterile Radiata Pine trees for regenerating softwood plantations after the final clear-fell harvest. This is because Radiata Pine wildlings threaten the integrity of patches of remnant native vegetation. Where possible, existing Radiata Pine wildlings should be removed from remnant vegetation patches.

The maintenance of biological legacies

As outlined in Chapter 9, biological legacies are organisms, organically derived structures and organically produced patterns that survive from the pre-disturbance system. Research on landscape genetics (see Chapter 9) and landscape context effects (see Chapter 5) demonstrated the importance of biological legacies for biotic persistence following disturbance at Tumut. For example, surveys of pine stands indicated that large eucalypt logs were important biological legacies. These logs are the remains of the original native forest that was cleared to establish pine stands. They provide habitat for a range of native animals (e.g. reptiles and beetles; see Fischer et al., 2005; Schmuki et al., 2006a) as well as substrates for plant species such as bryophytes (Pharo et al., 2004). Many species would not occur in pine stands in the absence of these key biological legacies, and the magnitude of negative landscape context effects is mitigated by their presence (Pharo et al., 2004; Fischer et al., 2005).

Based on the work highlighting the importance of biological legacies at Tumut, a general recommendation for plantation managers is that the decay of windrowed eucalypt logs should not be accelerated by damage from harvesting machinery during logging operations (Fischer *et al.*, 2005). The loss of these critical biological legacies would, for example, result in a reduction of up to 90% of the bryophytes that would otherwise occur in stands of Radiata Pine (Pharo *et al.*, 2004).

The lack of indicator species

The work at Tumut has clearly demonstrated that there is no viable indicator species that can act as a valid biodiversity surrogate. Much has been written on the identification of surrogates of biodiversity, in particular the selection of indicator species (Landres *et al.*, 1988; Lindenmayer *et al.*, 2000a; Maes and Van Dyck, 2005). The concept is a noble one in that if a species or small set of species could be identified that adequately reflect the condition of an ecosystem or the presence of the full suite of other species in an assemblage, then ecological assessments would simply have to focus on those indicator species.

The work at Tumut has demonstrated strongly contrasting landscape context effects among species as well as other kinds of responses to landscape condition (Lindenmayer *et al.*, 2002a; Chapters 5 and 7). Work at Nanangroe has produced similar outcomes, including many that were not readily predictable (Lindenmayer *et al.*, 2008b; Chapter 10). Although patterns of significant nestedness were identified for birds, they were not apparent for other vertebrate groups such as reptiles and mammals (Fischer and Lindenmayer, 2005a).

A clear conclusion from the body of work at Tumut is that there are no clear individual species or set of species that would be a useful indicator of the response of the rest of the biota, landscape condition or some other environment attribute of management interest. Rather than relying on the selection of a so-called 'indicator species' of dubious validity at best, a better approach would be to focus on the direct measurement of the entity of management interest – without assuming that it is a surrogate for other entities. This requires land managers to be explicit in their management objectives, such as which species or which states of landscape condition are the ones that need to be managed.

Summary

This chapter has outlined some recommendations for maintaining and/or enhancing the conservation value of plantations for biodiversity. The

Table 11.1. *Summary of recommendations for improved biodiversity conservation in plantation landscapes.*

Recommendation	Description	Reference in this book	Citation(s)
Retain patches of remnant native vegetation	Patches of native vegetation are used by a range of species in different groups	Chapters 4, 5	Lindenmayer *et al.* (1999a, 1999c, 2002a, 2002b)
Retain strips of native vegetation in gully lines	Native vegetation in gully lines provides a dispersal route for small mammals	Chapter 9	Lindenmayer and Peakall (2000); Banks *et al.* (2005a, 2005b, 2005c).
Retain the structure of vegetation in and around water bodies during the establishment of plantations	Aquatic environments are critical areas for amphibians	Chapter 5	Parris and Lindenmayer (2004)
Stagger clear-felling of pine stands around patches of remnant native vegetation	The level of patch–matrix contrast has a significant influence on bird and mammal assemblages	Chapter 10	Lindenmayer *et al.* (2008a)
Maintain biological legacies within pine stands	Biological legacies such as large eucalypt logs are valuable habitats for many species in pine stands	Chapter 5	Pharo *et al.* (2004), Fischer *et al.* (2005), Schmuki *et al.* (2006a)
Instigate hygiene protocols for logging machinery used in plantations	Invasive species such as Blackberry can be spread by machinery in existing and new plantations	Chapter 5	Lindenmayer and McCarthy (2001)

recommendations are based on the findings from Tumut as well as those from the work at Nanangroe (see Chapter 10). Many of these recommendations are relevant at the stand and landscape levels, as well as in newly established plantations and also in plantations that have already been established.

A particularly important finding was the value for a wide range of species of native forest remnants embedded within the plantation. The value of native vegetation for biota within plantations is a result consistent across studies from around the world (e.g. Peterken and Ratcliffe, 1995; Zanuncio *et al.*, 1998; Estades and Temple, 1999; Lindenmayer and Hobbs, 2004). This emphasises the critical importance of well-informed plantation design based not on the development of large-scale monocultures of introduced trees, but rather the creation of heterogeneous landscape mosaics (*sensu* Bennett *et al.*, 2006) where stands of planted trees are interspersed with patches of remnant native vegetation. Table 11.1 presents a summary of recommendations for improved biodiversity conservation in plantation landscapes.

As outlined in the following chapter, the work at Tumut is ongoing (Chapter 12) and new insights from it may result in additional recommendations for biodiversity conservation in plantations.

12 · *Lessons on running large-scale research studies: some insights from running the Tumut Fragmentation Study and directions for the future*

Before bringing this book to a close it is worth reflecting on some of the major issues and challenges associated with maintaining a large-scale, long-term project like the Tumut Fragmentation Study. Of course, much work remains to be done and the last part of this chapter discusses some directions for future work.

The challenges of maintaining a large-scale, multifaceted research project

It's my hope that this book has highlighted the diverse and multifaceted nature of the research programme undertaken at Tumut. Most of my colleagues consider that the work at Tumut has been productive and useful, and has changed thinking in some areas of ecology as well as altering approaches to the management of plantations. However, the maintenance of the work undertaken over the past 12+ years has involved more than its fair share of challenges. Three of these are outlined below. They are briefly summarised to illustrate some of the hurdles that can arise and that might hopefully be avoided by others embarking on long-term, large-scale research studies.

Data curation and management

The maintenance of a high-quality dataset is a pivotal part of any successful long-term ecological project. This sounds like a trivial point to most scientists. But the reality is that the curation of data is often an afterthought in the vast majority of ecological projects. Indeed, I have personally witnessed many cases where government agencies have discarded high-quality datasets that were not long afterwards recognised as being extremely important.

Over the past 5 years, a part-time database specialist has been employed to manage the data gathered from the various studies at Tumut (as well as data gathered from five other large-scale studies being conducted in south-eastern Australia). Data are added on a 6-monthly basis as additional field surveys are completed. The input of data requires careful checking, as error rates otherwise average 7–10% of the total input information.

Data storage remains a serious problem. For example, over the duration of the project at Tumut, the 'evolution' of data storage methods has included magnetic tapes, 5½ inch floppy disks, 3½ inch floppy disks, CDs, DVDs and memory sticks. The time to redundancy of these various storage methods has been steadily reducing. The development of new data storage methods has required regular updating, a process that can lead to errors being introduced to datasets or to data loss.

A key revelation has been the importance of primary data – in the form of pencil entries on paper copies of field datasheets. Many errors in digital datasets were rectified only by cross-reference to the original field data-sheets. The orderly storage of vast quantities of paper records is important, as are their duplication and storage off-site in case of an unforeseen event such as an office fire – a fate not unknown in some Australian government agencies.

The maintenance of a field presence

The various studies at Tumut have, in general, been characterised by the collection of high-quality empirical data gathered through repeated surveys from many permanently established field sites. The amount of field-work involved in these studies is considerable. For example, the completion of the cross-sectional study of landscape context effects at Tumut involved many visits to each of 166 field sites over many years (see Chapters 4 and 5); similarly for the 138 sites which underpin the project at Nanangroe (Chapter 10).

One of the lessons from running these projects has been that the most effective way to maintain them is to station a person permanently within or close to the study region – in this case, the town of Tumut. This has been the only way to complete the extensive field sampling associated with these large studies. This approach also has the advantage of ensuring close interactions with staff from the various agencies responsible for land management in the area. This has facilitated the rapid communication of results and can highlight the relevance of science to resource management. Regular contact with agency staff has become increasingly important in recent years because of high rates of staff turnover within those agencies.

Funding

Maintaining an ongoing field presence and constantly upgrading databases requires high-quality staff. This requires considerable amounts of money. Securing research funds is the bane of virtually all researchers' lives. Generating the funds to maintain the work at Tumut has been no exception. More than 40 grant proposals to over 15 different funding organisations have been submitted over the past 10 years; most were unsuccessful. Funding becomes more difficult to obtain as a large project progresses. This is because the emphasis of most funding bodies is on 'new work'. While new questions can be addressed in a location where research has been ongoing for some time, few funding bodies are willing to support long-term work that has been focused on the same area for a prolonged period (but see Laurance *et al.*, 2001; Haynes *et al.*, 2006; Stouffer *et al.*, 2006; for rare exceptions). This is unfortunate because the most interesting findings often emerge only when a project has been running for a considerable period of time.

One of the approaches to tackling the problems associated with the need for new research by funding bodies has been to use the major cross-sectional study comprising 166 sites at Tumut as a broad framework from which to 'hang' many other kinds of studies, including focused postgraduate research programmes. This has the advantage that true value-adding can be generated when insights from several strands of research can be integrated. An example from Tumut was the new insights into small mammal ecology that were derived from integrating empirical field data, molecular genetics and simulation modelling (see Chapters 5, 8 and 9).

Nevertheless, difficulty in obtaining ongoing funding remains the most significant threat to the maintenance of the work at Tumut. In the unfortunate event that the research programme was to be terminated through a shortage of funds, strict attention to data curation and database management aim to ensure that field information is stored in a way that allows the work to be recommenced by another researcher in the future if funding should again become available.

The Tumut and Nanangroe study areas are part of the Australian Long-term Ecological Research (LTER) network – an informal alliance of seven forest research sites nationwide. An aim of nominating the Tumut area for LTER status was to instigate succession planning so that another researcher could take over the work in the future. This is a pivotal point, as the vast majority of long-term studies are terminated when their 'champion' is no longer involved (Norton, 1996; Lindenmayer and Franklin, 2002).

Deficiencies in existing work and future work

In Chapter 10, problems with the cross-sectional nature of the main study of landscape context effects were discussed. This, in turn, triggered the work at Nanangroe. Issues were also raised with the rapid and substantial changes in the pine stands at Tumut and how the Edge Experiment was established to examine such changes. There are other problems with the work at Tumut (and indeed readers may identify others not outlined here). One of these has been the bias in pattern-based research and the limited emphasis on process-based work. The research at Tumut has been dominated by pattern-based studies such as those on the patterns of patch occupancy and patch-level animal abundance (Chapter 5), the prevalence of nestedness in species assemblages (Chapter 7) and the search for links between animal occurrence and landscape metrics (Chapter 6).

Pattern-based research dominates studies of landscape change and habitat fragmentation; often one pattern (e.g. the spatial distribution of a species or set of species) is correlated with another pattern (e.g. the spatial distribution of patches of native vegetation), with a limited ability to tease apart the ecological processes linking the two patterns (Lindenmayer and Fischer, 2007). However, without a focus on the processes or mechanisms giving rise to emergent patterns, much of the existing 'habitat fragmentation' literature has become dominated by the 'story-telling' of case studies describing species-specific or landscape-specific findings, which are difficult to generalise to other taxa or locations, thus thwarting attempts to identify 'clear insights into system dynamics' (Lindenmayer and Fischer, 2007).

More recent research projects at Tumut have attempted to better understand the ecological processes which have given rise to emergent patterns of animal presence, occurrence and assemblage composition. Examples include the integration of genetic and demographic research on the population recovery of the Bush Rat (Lindenmayer et al., 2005a; see

Chapter 9), connectivity and dispersal in the Greater Glider (Taylor *et al.*, 2007; see Chapter 9) and quantification of how patches are used by arboreal marsupials and birds (Pope *et al.*, 2004; Tubelis *et al.*, 2004, 2007a; see Chapter 6). These projects are a useful start, but it is well recognised that much more process-based work is needed, especially as the patterns of animal distribution and abundance are very well documented and form an excellent platform for well-directed process-oriented work.

Additional process-based work is planned that will quantify source–sink dynamics (*sensu* Pulliam *et al.*, 1992) for small mammal populations and how this might contribute to the patterns of patch occupancy that have been observed at Tumut (Banks, 2005). Another study is one on foliage nutrient levels in eucalypt trees within patches of remnant native vegetation and the eucalypt controls (Youngentob *et al.*, 2008). Here the aim is to determine whether there are links between the quality of food resources for arboreal marsupials, landscape context and patterns of site occupancy. In both cases, one aim is to examine potential ecological mechanisms or processes that might help to better explain patterns of animal presence and abundance that have been observed at Tumut.

Of course, an increased emphasis on process-based work at Tumut does not mean that it is appropriate to dispense with pattern-based work entirely. There are important synergies between pattern- and process-based studies (Lindenmayer and Fischer, 2006, 2007; Lindenmayer *et al.*, 2007b) which necessitate maintaining both kinds of work. For example, part of the ongoing work on arboreal marsupials at Tumut will involve recounting animals in the eucalypt controls as well as in the remnant patches, and comparing findings with earlier surveys (completed between 1996 and 1999). Extensive clear-felling of pine stands at Tumut will make it possible to further document landscape context effects for the patches of remnant eucalypt forest (K. Youngentob *et al.*, unpublished).

Other important work remains to be done at Tumut. Opportunities exist to revisit the sites used in the removal experiment of small mammals in patches of remnant eucalypt forest (Lindenmayer *et al.*, 2005a) to determine whether the patch extinctions created in a few areas have been reversed.

Finally, a strength of the work at Nanangroe lies with repeated surveys over a prolonged period, given that the work was designed as a true longitudinal natural experiment (Lindenmayer *et al.*, 2008b). Therefore, recounts of birds, mammals and reptiles will continue there for at least the next 10–15 years, when the system will reach the end of the first rotation and pine stands will be clear-felled. In addition, vegetation structure and plant species composition will continue to be measured at Nanangroe so that changes in vertebrate populations can be linked with vegetation changes.

Some concluding remarks

The research programme at Tumut has been a challenging and exciting journey over the past 13 years. The work has expanded from a single large-scale landscape context study (Chapters 4 and 5) to encompass a far wider range of investigations than could have been envisaged in 1995. The work has been built on a close collaborative partnership between ecologists, statisticians, modellers and geneticists.

In particular, the role of expert professional statisticians has been prominent, particularly in contributing to experimental design, data analysis and interpretation. Such partnerships form the basis of current ongoing projects at Tumut and nearby Nanangroe and would undoubtedly be the foundation of any future study that may take place in the region. They have facilitated the development of highly productive research programmes that have generated new insights far beyond what would have been possible from the efforts of workers in any one discipline alone.

My hope is that the Tumut Fragmentation Study, as a model of what a collaborative partnership can achieve, might inspire other workers to establish research programmes of similar scope. In so doing, we might be better placed to understand the biotic effects of landscape modification. And that should empower us to more effectively conserve biodiversity.

Appendix 1: List of collaborators/contributors to the Tumut Fragmentation Experiment

Principal Investigators

Professor David Lindenmayer (Fenner School of Environment and Society, The Australian National University, Canberra).

Adjunct Professor Ross Cunningham (Fenner School of Environment and Society, The Australian National University, Canberra).

Field Research Staff

Mr Mason Crane (2000 onwards) (Fenner School of Environment and Society, The Australian National University, Canberra).

Mr Ryan Incoll (1996–2000) (Fenner School of Environment and Society, The Australian National University, Canberra).

Mr Chris MacGregor (1997 onwards) (Fenner School of Environment and Society, The Australian National University, Canberra).

Mr Lachlan McBurney (2006 onwards) (Fenner School of Environment and Society, The Australian National University, Canberra).

Mr Damian Michael (2002 onwards) (Fenner School of Environment and Society, The Australian National University, Canberra).

Dr Rebecca Montague-Drake (2006 onwards) (Fenner School of Environment and Society, The Australian National University, Canberra).

Mr Matthew Pope (1995–1998) (Fenner School of Environment and Society, The Australian National University, Canberra).

Mr David Rawlins (2003) (Fenner School of Environment and Society, The Australian National University, Canberra).

Mr Craig Tribolet (1998–2000) (Fenner School of Environment and Society, The Australian National University, Canberra).

Scientific Investigators and collaborators (alphabetical order)

Dr Sam Banks (Ecologist and Geneticist) (Fenner School of Environment and Society, The Australian National University, Canberra).

Dr Andrew Claridge (Ecologist) (NSW National Parks and Wildlife Service, Queanbeyan, NSW).

Ms Christine Donnelly (Data Analyst) (Statistical Consulting Unit, The Australian National University, Canberra).

Dr Don Driscoll (Ecologist) (Fenner School of Environment and Society, The Australian National University, Canberra).

Dr Adam Felton (Ecologist) (Fenner School of Environment and Society, The Australian National University, Canberra).

Dr Joern Fischer (Ecologist) (Fenner School of Environment and Society, The Australian National University, Canberra).

Dr Emma Knight (Statistician) (Fenner School of Environment and Society, The Australian National University, Canberra).

Dr Robert Lacy, (Mathematical Modeller and Geneticist) (Center for Conservation Biology, Brookfield Zoo, Chicago, USA).

Dr Sarah Legge (Behavioural Ecologist) (Division of Botany and Zoology, The Australian National University, Canberra).

Dr Rob Lesslie (Ecologist) (Department of Geography, The Australian National University, Canberra).

Dr Michael McCarthy (Mathematical Modeller and Forest Ecologist) (Australian Research Centre for Urban Ecology, The University of Melbourne, Melbourne).

Mr Ed Merrett (GIS Unit, State Forests of NSW, Albury, NSW).

Professor Henry Nix (Ecologist and Climate Analysis Expert) (Fenner School of Environment and Society, The Australian National University, Canberra).

Dr Kirsten Parris (Herpetologist and Ecologist) (School of Botany, The University of Melbourne, Melbourne).

Professor Rod Peakall (Geneticist) (Division of Botany and Zoology, The Australian National University, Canberra).

Professor Hugh Possingham (Mathematical Modeller and Ecologist) (The Ecology Centre, The University of Queensland, Brisbane).

Ms Monica Ruibal (Geneticist) (Division of Botany and Zoology, The Australian National University, Canberra).

Dr Jeanette Stanley (Social Scientist) (Fenner School of Environment and Society, The Australian National University, Canberra).

Mr John Stein (Climate and GIS Analyst) (Fenner School of Environment and Society, The Australian National University, Canberra).

Dr Paul Sunnucks (Geneticist) (Department of Environmental Science, Monash University, Clayton, Victoria).

Dr Andrea Taylor (Geneticist) (Department of Environmental Science, Monash University, Clayton, Victoria).

Professor Hugh Tyndale-Biscoe (Ecologist) (Fenner School of Environment and Society, The Australian National University, Canberra).

Dr Karen Viggers (Veterinary Scientist and Ecologist) (Research School of Biological Sciences, The Australian National University, Canberra).

Professor Alan Welsh (Statistician) (Centre for Mathematics and its Applications, The Australian National University, Canberra).

Associate Professor Jeff Wood (Statistician) (Fenner School of Environment and Society, The Australian National University, Canberra).

Appendix 2: Detections of bird species in the Tumut Fragmentation Study classified by four broad classes of sites

Common name	Scientific name	Eucalypt controls	Eucalypt patches	Eucalypt strips	Radiata Pine
FAMILY: ACCIPITRIDAE					
Brown Goshawk	*Accipiter fasciatus*	Rare	Rare	Rare	Absent
Collared Sparrowhawk	*Accipiter cirrhocephalus*	Rare	Absent	Absent	Absent
Little Eagle	*Hieraaetus morphnoides*	Absent	Rare	Rare	Absent
Wedge–tailed Eagle	*Aquila audax*	Rare	Rare	Rare	Absent
FAMILY: ANATIDAE					
Pacific Black Duck	*Anas superciliosa*	Absent	Absent	Rare	Rare
Wood Duck	*Chenonetta jubata*	Rare	Absent	Rare	Absent
FAMILY: ARTAMIDAE					
Australian Magpie	*Gymnorhina tibicen*	Rare	Rare	Rare	Rare
Dusky Woodswallow	*Artamus cyanopterus*	Rare	Absent	Rare	Absent
Grey Butcherbird	*Cracticus torquatus*	Rare	Rare	Rare	Rare
Grey Currawong	*Strepera versicolor*	Rare	Rare	Rare	Absent
Pied Currawong	*Strepera graculina*	Sparse	Sparse	Rare	Sparse
FAMILY: CACATUIDAE					
Galah	*Cacatua roseicapilla*	Rare	Rare	Rare	Absent
Gang–gang Cockatoo	*Callocephalon fimbriatum*	Rare	Rare	Rare	Rare
Sulphur–crested Cockatoo	*Cacatua galerita*	Common	Common	Sparse	Sparse
Yellow–tailed Black Cockatoo	*Calyptorhynchus funereus*	Rare	Rare	Rare	Rare
FAMILY: CAMPEPHAGIDAE					
Black–faced Cuckoo Shrike	*Coracina novaehollandiae*	Rare	Rare	Rare	Rare
Cicadabird	*Coracina tenuirostris*	Rare	Absent	Absent	Absent
White–winged Triller	*Lalage sueurii*	Absent	Absent	Rare	Absent

Family / Species	Scientific name					
FAMILY: CHARADRIIDAE						
Masked Lapwing	*Vanellus miles*	Absent	Rare	Rare	Rare	Absent
FAMILY: CINCLOSOMATIDAE						
Eastern Whipbird	*Psophodes olivaceus*	Rare	Rare	Rare	Rare	Rare
Spotted Quail Thrush	*Cinclosoma punctatum*	Rare	Absent	Absent	Absent	Rare
FAMILY: CLIMACTERIDAE						
Red-browed Treecreeper	*Climacteris erythrops*	Rare	Rare	Rare	Rare	Absent
White-throated Treecreeper	*Cormobates leucophaeus*	Common	Common	Common	Common	Rare
FAMILY: COLUMBIDAE						
Common Bronzewing	*Phaps chalcoptera*	Rare	Rare	Rare	Rare	Rare
Peaceful Dove	*Geopelia striata*	Absent	Absent	Absent	Absent	Rare
Wonga Pigeon	*Leucosarcia melanoleuca*	Rare	Rare	Rare	Rare	Absent
FAMILY: CORCORACIDAE						
White-winged Chough	*Corcorax melanorhamphos*	Rare	Rare	Rare	Rare	Rare
FAMILY: CORVIDAE						
Australian Raven	*Corvus coronoides*	Rare	Rare	Rare	Rare	Rare
Little Raven	*Corvus mellori*	Rare	Rare	Rare	Rare	Rare
FAMILY: CUCULIDAE						
Brush Cuckoo	*Cacomantis variolosus*	Rare	Absent	Absent	Rare	Rare
Fan-tailed Cuckoo	*Cacomantis flabelliformis*	Rare	Rare	Rare	Rare	Rare
Horsfield's Bronze Cuckoo	*Chrysococcyx basilis*	Rare	Rare	Rare	Rare	Rare
Pallid Cuckoo	*Cuculus pallidus*	Absent	Absent	Absent	Absent	Absent
Shining Bronze Cuckoo	*Chrysococcyx lucidus*	Rare	Rare	Rare	Rare	Rare
FAMILY: DICAEIDAE						
Mistletoebird	*Dicaeum hirundinaceum*	Rare	Rare	Rare	Rare	Rare

(cont.)

Common name	Scientific name	Eucalypt controls	Eucalypt patches	Eucalypt strips	Radiata Pine
FAMILY: DICRUDIDAE					
Grey Fantail	*Rhipidura fuliginosa*	Common	Common	Abundant	Sparse
Leaden Flycatcher	*Myiagra rubecula*	Rare	Rare	Rare	Rare
Magpie Lark	*Grallina cyanoleuca*	Absent	Rare	Absent	Rare
Rufous Fantail	*Rhipidura rufifrons*	Rare	Rare	Rare	Rare
Satin Flycatcher	*Myiagra cyanoleuca*	Rare	Rare	Rare	Absent
Willie Wagtail	*Rhipidura leucophrys*	Absent	Rare	Rare	Rare
FAMILY: FALCONIDAE					
Brown Falcon	*Falco berigora*	Absent	Rare	Absent	Absent
Nankeen Kestrel	*Falco cenchroides*	Absent	Rare	Rare	Absent
FAMILY: FRINGILLIDAE					
European Goldfinch	*Carduelis carduelis*	Rare	Rare	Rare	Rare
FAMILY: HALCYONIDAE					
Laughing Kookaburra	*Dacelo novaeguineae*	Rare	Rare	Rare	Rare
Sacred Kingfisher	*Todiramphus sanctus*	Rare	Rare	Rare	Rare
FAMILY: HIRUNDINIDAE					
Tree Martin	*Hirundo nigricans*	Absent	Rare	Absent	Absent
Welcome Swallow	*Hirundo neoxena*	Absent	Absent	Rare	Absent
FAMILY: MALURIDAE					
Superb Fairy Wren	*Malurus cyaneus*	Rare	Rare	Rare	Rare

FAMILY / Common name	Scientific name				
FAMILY: MELIPHAGIDAE					
Brown-headed Honeyeater	*Melithreptus brevirostris*	Rare	Rare	Rare	Rare
Crescent Honeyeater	*Phylidonyris pyrrhoptera*	Rare	Rare	Rare	Rare
Eastern Spinebill	*Acanthorhynchus tenuirostris*	Rare	Rare	Rare	Rare
Noisy Friarbird	*Philemon corniculatus*	Rare	Rare	Rare	Rare
Red Wattlebird	*Anthochaera carunculata*	Common	Rare	Rare	Rare
White-eared Honeyeater	*Lichenostomus leucotis*	Rare	Rare	Rare	Rare
White-naped Honeyeater	*Melithreptus lunatus*	Sparse	Rare	Rare	Rare
Yellow-faced Honeyeater	*Lichenostomus chrysops*	Abundant	Abundant	Abundant	Sparse
FAMILY: MENURIDAE					
Superb Lyrebird	*Menura novaehollandiae*	Rare	Rare	Rare	Rare
FAMILY: MEROPIDAE					
Rainbow Bee-eater	*Merops ornatus*	Absent	Absent	Rare	Absent
FAMILY: MUSCICAPIDAE					
Bassian Ground Thrush	*Zoothera lunulata*	Rare	Rare	Rare	Rare
Blackbird	*Turdus merula*	Rare	Rare	Rare	Sparse
FAMILY: NEOSITTIDAE					
Varied Sittella	*Daphoenositta chrysoptera*	Rare	Rare	Rare	Rare
FAMILY: ORIOLIDAE					
Olive-backed Oriole	*Oriolus sagittatus*	Rare	Absent	Rare	Rare
FAMILY: PACHYCEPHALIDAE					
Crested Shrike Tit	*Falcunculus frontatus*	Rare	Rare	Rare	Rare
Golden Whistler	*Pachycephala pectoralis*	Rare	Rare	Sparse	Sparse
Grey Shrike Thrush	*Colluricincla harmonica*	Common	Common	Common	Common
Olive Whistler	*Pachycephala olivacea*	Rare	Rare	Rare	Rare
Rufous Whistler	*Pachycephala rufiventris*	Sparse	Common	Common	Common

(*cont.*)

Common name	Scientific name	Eucalypt controls	Eucalypt patches	Eucalypt strips	Radiata Pine
FAMILY: PARDALOTIDAE					
Brown Thornbill	*Acanthiza pusilla*	Sparse	Sparse	Sparse	Sparse
Buff-rumped Thornbill	*Acanthiza reguloides*	Rare	Rare	Absent	Rare
Pilotbird	*Pynoptilus floccosus*	Rare	Rare	Rare	Rare
Spotted Pardalote	*Pardalotus punctatus*	Common	Sparse	Sparse	Rare
Striated Pardalote	*Pardalotus striatus*	Common	Sparse	Rare	Rare
Striated Thornbill	*Acanthiza lineata*	Rare	Rare	Rare	Rare
Western Gerygone	*Gerygone fusca*	Rare	Absent	Absent	Absent
White-browed Scrub Wren	*Sericornis frontalis*	Sparse	Common	Common	Abundant
White-throated Gerygone	*Gerygone olivacea*	Rare	Rare	Rare	Absent
Yellow Thornbill	*Acanthiza nana*	Rare	Absent	Absent	Rare
Yellow-rumped Thornbill	*Acanthiza chrysorrhoa*	Absent	Rare	Rare	Absent
FAMILY: PASSERIDAE					
Red-browed Finch	*Neochmia temporalis*	Rare	Rare	Rare	Rare
FAMILY: PETROICIDAE					
Eastern Yellow Robin	*Eopsaltria australis*	Rare	Sparse	Sparse	Sparse
Flame Robin	*Petroica phoenicea*	Rare	Rare	Rare	Rare
Red-capped Robin	*Petroica goodenovii*	Absent	Absent	Absent	Rare
Rose Robin	*Petroica rosea*	Rare	Rare	Rare	Rare
Scarlet Robin	*Petroica multicolor*	Rare	Rare	Rare	Rare
FAMILY: PHALACROCORACIDAE					
Little Black Cormorant	*Phalacrocorax sulcirostris*	Rare	Absent	Absent	Absent

FAMILY: PSITTACIDAE					
Crimson Rosella	*Platycercus elegans*	Sparse	Common	Common	Sparse
Eastern Rosella	*Platycercus eximius*	Absent	Rare	Absent	Absent
King Parrot	*Alisterus scapularis*	Rare	Rare	Rare	Rare
FAMILY: PTILONORHYNCHIDAE					
Satin Bowerbird	*Ptilonorhychus violaceus*	Rare	Rare	Rare	Rare
FAMILY: STURNIDAE					
Common Starling	*Sturnus vulgaris*	Absent	Rare	Rare	Absent
FAMILY: ZOSTEROPIDAE					
Silvereye	*Zosterops lateralis*	Rare	Rare	Rare	Sparse

Codes are: Rare (detected at <25% of sites), uncommon (detected at 25–50% of sites), common (detected at 50–75% of sites) and abundant (detected at >75% of sites) (modified from Lindenmayer et al., 2007c).

References

Alatalo, R.V., Lundberg, A. and Bjorklund, M. (1982). Can the song of male birds attract other males? An experiment with the pied flycatcher *Ficedula hypoleuca*. *Bird Behaviour*, **4**, 42–45.

Allen, R., Platt, K. and Wiser, S. (1995). Biodiversity in New Zealand plantations. *New Zealand Forestry*, **39**, 26–29.

Andrén, H. (1994). Effects of habitat fragmentation on birds and mammals in landscapes with different proportions of suitable habitat: a review. *Oikos*, **71**, 355–366.

Andrén, H. (1997). Habitat fragmentation and changes in biodiversity. *Ecological Bulletin*, **46**, 171–181.

Andrén, H. (1999). Habitat fragmentation, the random sample hypothesis and critical thresholds. *Oikos*, **84**, 306–308.

Angermeier, P.I. (1995). Ecological attributes of extinction-prone species: loss of freshwater fishes of Virginia. *Conservation Biology*, **9**, 143–158.

Armbruster, P. and Lande, R. (1993). A population viability analysis for African elephant (*Luxodonta africana*): how big should reserves be? *Conservation Biology*, **7**, 602–610.

Arnold, G.W., Steven, D.E. and Weeldenburg, J.R. (1993). Influences of remnant size, spacing pattern and connectivity on population boundaries and demography in euros *Macropus robustus* living in a fragmented landscape. *Biological Conservation*, **64**, 219–230.

Arrhenius, O. (1921). Species and area. *Journal of Ecology*, **9**, 95–99.

Atmar, W. and Patterson, B.D. (1995). *The Nestedness Temperature Calculator: A Visual Basic Program, including 294 Presence–Absence Matrices.* Chicago IL: AICS Research Inc.

Austin, M.P. (1999). A silent clash of paradigms: some inconsistencies in community ecology. *Oikos*, **86**, 170–178.

Ball, S., Lindenmayer, D.B. and Possingham, H.P. (2003). The predictive accuracy of viability analysis: a test using data from two small mammal species in a fragmented landscape. *Biodiversity and Conservation*, **12**, 2393–2413.

Banks, S.C. (2005). Habitat fragmentation impacts on population processes in *Antechinus agilis*. Unpublished Ph.D. thesis, Monash University, Melbourne, Australia.

Banks, S.C., Lindenmayer, D.B., Ward, S.J. and Taylor, A.C. (2005a). The effects of habitat fragmentation via forestry plantation establishment on spatial geno-type structure in a small marsupial carnivore, *Antechinus agilis*. *Molecular Ecology*, **14**, 1667–1680.

Banks, S.C., Ward, S.J., Lindenmayer, D.B., Finlayson, G.R., Lawson, S.J., and Taylor, A.C. (2005b). The effects of habitat fragmentation on the social kin structure and mating system of the agile antechinus *Antechinus agilis*. *Molecular Ecology*, **14**, 1789–1801.

Banks, S.C., Lawson, S.J., Finlayson, G.R., Lindenmayer, D.B., Ward, S.J. and Taylor, A.C. (2005c). The effects of habitat fragmentation on demography and genetic variation of a marsupial carnivore, *Antechinus agilis*. *Biological Conservation*, **122**, 581–597.

Banks, S.C., Piggott, M.P., Stow, A.J. and Taylor, A.C. (2007). Sex and sociality in a disconnected world: a review of the impacts of habitat fragmentation on animal social interactions. *Canadian Journal of Zoology*, **85**, 1065–1079.

Barlow, J., Gardner, T.A., Araujo, I.S., Ávila-Pires, T.C., Bonaldo, A.B., Costa, J.E., Esposito, M.C., Ferreira, L.V., Hawes, J., Hernandez, M.I.M., Hoogmoed, M.S., Leite, R.N., Lo-Man-Hung, N.F., Malcolm, J.R., Martins, M.B., Mestre, L.A.M., Miranda-Santos, R., Nunes-Gutjahr, A.L., Overal, W.L., Parry, L., Peters, S.L., Ribeiro-Junior, M.A., da Silva, M.N.F., da Silva Motta, C. and Peres, C.A. (2007). Quantifying the biodiversity value of tropical primary, secondary, and plantation forests. *Proceedings of the National Academy of Sciences*, **104**, 18555–18560.

Barker, J., Grigg, G.C. and Tyler, M.J. (1995). *A Field Guide to Australian Frogs*. Chipping Norton, NSW, Australia: Surrey Beatty and Sons.

Barrett, G. (2000). Birds on farms: ecological management for agricultural sustainability. *Wingspan*, **10**(4) (Suppl.), 1–16.

Barrett, G.W., Ford, H.A. and Recher. H.F. (1994). Conservation of woodland birds in a fragmented rural landscape. *Pacific Conservation Biology*, **1**, 245–256.

Barrett, G., Silcocks, A., Barry, S., Cunningham, R. and Poulter, R. (2003). *New Atlas of Australian Birds*. Melbourne, Australia: Birds Australia.

Bayne, E.M. and Hobson, K.A. (1997). Comparing the effects of landscape fragmentation by forestry and agriculture on predation of artificial nests. *Conservation Biology*, **11**, 1418–1429.

Bayne, E.M. and Hobson, K.A. (1998). The effects of habitat fragmentation by forestry and agriculture on the abundance of small mammals in the southern boreal mixedwood forest. *Canadian Journal of Zoology*, **76**, 62–69.

Beier, P. and Noss, R. (1998). Do habitat corridors provide connectivity? *Conservation Biology*, **12**, 1241–1252.

Beissinger, S.R., and McCullough, D.R. (2002). *Population Viability Analysis*. Chicago, IL: University of Chicago Press.

Bender, D.J., and Fahrig, L. (2005). Matrix structure obscures the relationship between interpatch movement and patch size and isolation. *Ecology*, **86**, 1023–1033.

Bender, D.J., Contreras, T.A. and Fahrig, L. (1998). Habitat loss and population decline: a meta-analysis of the patch size effect. *Ecology*, **79**, 517–529.

Bender, D.J., Tischendorf, L. and Fahrig, L. (2003). Evaluation of patch isolation metrics for predicting animal movement in binary landscapes. *Landscape Ecology*, **18**, 17–39.

Bennett, A.F. (1990). *Habitat Corridors: Their Role in Wildlife Management and Conservation*. Melbourne, Australia: Department of Conservation and Environment.

Bennett, A.F. (1998). *Linkages in the Landscape: The Role of Corridors and Connectivity in Wildlife Conservation*. Gland, Switzerland: IUCN.

Bennett, A.F., Radford, J.Q. and Haslem, A. (2006). Properties of land mosaics: implications for nature conservation in agricultural landscapes. *Biological Conservation*, **133**, 250–264.

Berglund, H. and Jonsson, B.G. (2003). Nested plant and fungal communities; the importance of area and habitat quality in maximizing species capture in boreal old-growth forests. *Biological Conservation*, **112**, 319–328.

Berry, O., Tocher, M.D., Gleeson, D.M. and Sarre, S. (2005). Effect of vegetation matrix on animal dispersal: genetic evidence from a study of endangered skinks. *Conservation Biology*, **19**, 855–864.

Bierregaard, R.O., Gascon, C., Lovejoy, T.E. and Mesquita, R. (eds.) (2001). *Lessons from Amazonia: The Ecology and Conservation of a Fragmented Forest*. New Haven, CT: Yale University Press.

Blakers, M., Davies, S.J. and Reilly, P.N. (1984). *The Atlas of Australian Birds*. Melbourne, Australia: Melbourne University Press.

Block, W.M. and Brennan, L.A. (1993). The habitat concept in ornithology: theory and applications. *Current Ornithology*, **11**, 35–91.

Boecklen, W.J. (1997). Nestedness, biogeographic theory, and the design of nature reserves. *Oecologia*, **112**, 123–142.

Boyce, M.S. (1992). Population viability analysis. *Annual Review of Ecology and Systematics*, **23**, 481–506.

Bradford, D.F., Neale, A.C., Nash, M.S., Sada, D.W. and Jaeger, J.R. (2003). Habitat patch occupancy by toads (*Bufo punctatus*) in a naturally fragmented desert landscape. *Ecology*, **84**, 1012–1023.

Bradstock, R.A., Bedward, M., Gill, A.M. and Cohn, J.S. (2005). Which mosaic? A landscape ecological approach for evaluating between fire regimes, habitat and animals. *Wildlife Research*, **32**, 409–423.

Brokaw, N.V. and Lent, R.A. (1999). Vertical structure. In *Managing Biodiversity in Forest Ecosystems*, ed. M. Hunter III. Cambridge, UK: Cambridge University Press, pp. 373–399.

Brook, B.W., O'Grady, J.J., Burgman, M.A., Akçakaya, H.R. and Frankham, R. (2000). Predictive accuracy of population viability analysis in conservation biology. *Nature*, **404**, 385–387.

Brook, B.W., Burgman, M.A., Akçakaya, H.R., O'Grady, J.J. and Frankham, R. (2002). Critiques of PVA ask the wrong questions: throwing out the heuristic baby with the numerical bath water. *Conservation Biology*, **16**, 262–263.

Brooker, L. and Brooker, M. (2002). Dispersal and population dynamics of the blue-breasted fairy-wren, *Malurus pulcherrimus*, in fragmented habitat in the Western Australian wheatbelt. *Wildlife Research*, **29**, 225–233.

Brothers, T.S. and Spingarn, A. (1992). Forest fragmentation and alien plant invasion of central Indiana old-growth forests. *Conservation Biology*, **6**, 91–100.

Brown, J.H. and Kodric-Brown, A. (1977). Turnover rates in insular biogeography: effect of immigration on extinction. *Ecology*, **58**, 445–449.

Brown, J.W. (1987). The peninsular effect in Baja California: an entomological assessment. *Journal of Biogeography*, **14**, 359–365.

Bunnell, F. (1999a). Foreword. Let's kill a panchreston: giving fragmentation a meaning. In *Forest Wildlife and Fragmentation: Management Implications*, ed. J. Rochelle, L.A. Lehmann and J. Wisniewski. Leiden, Germany: Brill, pp. vii–xiii.

Bunnell, F. (1999b). What habitat is an island? In *Forest Wildlife and Fragmentation: Management Implications*, ed. J. Rochelle, L. A. Lehmann and J. Wisniewski. Leiden, Germany: Brill, pp. 1–31.

Bunnell, F., Dunsworth, G., Huggard, D. and Kremsater, L. (2003). *Learning to Sustain Biological Diversity on Weyerhauser's Coastal Tenure*. Vancouver, BC, Canada: Weyerhauser Company.

Burdon, J. J. and Chilvers, G. A. (1994). Demographic changes and the development of competition in a native eucalypt forest invaded by exotic pines. *Oecologia*, **97**, 419–423.

Burgman, M., Ferson, S. and Akçakaya, H. R. (1993). *Risk Assessment in Conservation Biology*. New York: Chapman and Hall.

Burgman, M. A., Lindenmayer, D. B. and Elith, J. (2005). Managing landscapes for conservation under uncertainty. *Ecology*, **86**, 2007–2017.

Burns, K., Walker, D. and Hansard, A. (1999). *Forest Plantations on Cleared Agricultural Land in Australia: A Regional and Economic Analysis*. Research Report No. 99/11. Canberra, Australia: Australian Bureau of Agricultural and Resource Economics.

Busack, S. D. and Hedges, S. B. (1984). Is the peninsular effect a red herring? *American Naturalist*, **123**, 266–275.

Carpenter, S., Chisholm, S. W., Krebs, C. J., Schindler, C. J. and Wright, R. F. (1995). Ecosystems experiments. *Science*, **269**, 324–327.

Cary, G. (2002). Importance of a changing climate for fire regimes in Australia. In *Flammable Australia: The Fire Regimes and Biodiversity of a Continent*, ed. R. A. Bradstock, J. E. Williams and A. M. Gill. Melbourne, Australia: Cambridge University Press, pp. 26–48.

Cascante, A., Quesada, M., Lobo, J. J. and Fuchs, E. A. (2002). Effects of dry tropical forest fragmentation on the reproductive success and genetic structure of the tree *Samanea saman*. *Conservation Biology*, **16**, 137–147.

Caughley, G. (1994). Directions in conservation biology. *Journal of Animal Ecology*, **63**, 215–244.

Caughley, J. and Gall, B. (1985). Relevance of zoological transition to conservation of fauna: amphibians and reptiles in the south-western slopes of New South Wales. *Australian Zoologist*, **21**, 513–527.

Chalfoun, A. D., Thompson, F. R. and Ratnaswamy, M. J. (2002). Nest predators and fragmentation: a review and meta-analysis. *Conservation Biology*, **16**, 306–318.

Chen, J. (1991). *Edge effects: microclimatic pattern and biological responses in old-growth Douglas-fir forests*. Unpublished Ph.D. thesis, University of Washington, Seattle, WA, USA.

Chen, J., Franklin, J. F. and Spies, T. A. (1990). Microclimatic pattern and basic biological responses at the clearcut edges of old-growth Douglas-fir stands. *Northwest Environmental Journal*, **6**, 424–425.

Claridge, A. W., Paull, D., Dawson, J., Mifsud, G., Murray, A. J., Poore, R. and Saxon, M. J. (2005). Home range of the spotted-tailed quoll (*Dasyurus maculatus*), a marsupial carnivore, in a rainshadow woodland. *Wildlife Research*, **32**, 7–14.

Clout, M. N., and Gaze, P. D. (1984). Effects of plantation forestry on birds in New Zealand. *Journal of Applied Ecology*, **21**, 795–815.

Cockburn, A., Scott, M. P. and Scotts, D. J. (1985). Inbreeding avoidance and male-biased natal dispersal in *Antechinus* spp. (Marsupialia: Dasyuridae). *Animal Behaviour*, **33**, 908–915.

Connell, J.H. (1978). Diversity in tropical forests and coral reefs. *Science*, **199**, 1302–1310.

Cook, R.E. (1969). Variation in species density of North American birds. *Systematic Zoology*, **18**, 63–84.

Cornelius, C., Cofre, H. and Marquet, P.A. (2000). Effects of habitat fragmentation on bird species in a relict temperate forest in semiarid Chile. *Conservation Biology*, **14**, 534–543.

Costermans, L. (1994). *Native Trees and Shrubs of South-eastern Australia*, 2nd edn. Sydney, Australia: Rigby.

Coulson, T., Mace, G.M., Hudson, E. and Possingham, H. (2001). the use and abuse of population viability analysis. *Trends in Ecology and Evolution*, **16**, 219–221.

Craig, S.A. (1985). Social organization, reproduction and feeding behaviour of a population of yellow-bellied gliders, *Petaurus australis* (Marsupialia: Petauridae). *Australian Wildlife Research*, **12**, 1–18.

Crow, J.F. and Kimura, M. (1970). *An Introduction to Population Genetics Theory*. New York: Harper and Row.

Cubbage, F.W., Dvorak, W.S., Abt, R.C. and Pacheco, G. (1996). *World Timber Supply and Prospects: Models, Projections, Plantations and Implications*. Central America and Mexico Coniferous (CAMCORE) Annual Meeting, Bali, Indonesia.

Cunningham, R.B., Lindenmayer, D.B., Nix, H.A. and Lindenmayer, B.D. (1999). Quantifying observer heterogeneity in bird counts. *Australian Journal of Ecology*, **24**, 270–277.

Cunningham, R.B., Pope, M.L. and Lindenmayer, D.B. (2004a). Patch use by the greater glider (*Petauroides volans*) in a fragmented forest ecosystem. III. Night-time use of trees. *Wildlife Research*, **31**, 579–585.

Cunningham, R.B., Lindenmayer, D.B. and Lindenmayer, B.D. (2004b). Sound recording of bird vocalisations in forests. I. Relationships between bird vocalisations and point interval counts of bird numbers – a case study in statistical modelling. *Wildlife Research*, **31**, 195–207.

Cunningham, R.B., Lindenmayer, D.B., MacGregor, C. and Barry, S. (2005). Small mammal populations in a wet eucalypt forest: factors influencing the probability of capture. *Wildlife Research*, **32**, 657–671.

Cunningham, R.B., Lindenmayer, D.B., Crane, M., Michael, D. and McGregor, C. (2007). Reptile and arboreal marsupial response to replanted vegetation in agricultural landscapes. *Ecological Applications*, **17**, 609–619.

Cunningham, R.B., Lindenmayer, D.B., McGregor, C., Crane, M. and Michael, D. (2008). What factors influence bird biota on farms? Putting restored vegetation into context? *Conservation Biology*, **22**, 742–752.

Cuthill, I.C. and MacDonald, W.A. (1990). Experimental manipulation of the dawn and dusk chorus in the blackbird *Turdus merula*. *Behavioral Ecology and Sociobiology*, **26**, 209–216.

Cutler, A. (1991). Nested faunas and extinction in fragmented habitats. *Conservation Biology*, **5**, 496–505.

Daily, G.C. (2001). Ecological forecasts. *Nature*, **411**, 245.

Darlington, P.J.J. (1957). *Zoogeography*. New York: John Wiley and Sons.

Davey, S.M. (1990). Methods for surveying the abundance and distribution of arboreal marsupials in a south coast forest of New South Wales. *Australian Wildlife Research*, **17**, 427–445.

Davies, K.F., Margules, C.R. and Lawrence, J.F. (2000). Which traits of species predict population declines in experimental forest fragments? *Ecology*, **81**, 1450–1461.

Davies, K.F., Melbourne, B.A. and Margules, C.R. (2001). Effects of within- and between-patch processes on community dynamics in a fragmentation experiment. *Ecology*, **82**, 1830–1846.

Davies, N.B. and Lundberg, A. (1984). Food distribution and a variable mating system in the dunnock (*Prunella modularis*). *Journal of Animal Ecology*, **53**, 895–912.

Dawson, T.J. (1995). *Kangaroos: Biology of the Largest Marsupials*. Sydney, Australia: University of New South Wales Press.

Debinski, D.M. and Holt, R.D. (2000). A survey and overview of habitat fragmentation experiments. *Conservation Biology*, **14**, 342–355.

Department of Environment, Heritage and the Arts (2008). *Threatened Species and Threatened Ecological Communities*. Canberra, Australia: Department of Environment, Heritage and the Arts [available at www.environment.gov.au/biodiversity/threatened/index.html].

Diamond, J.M. (1985). How many unknown species are yet to be discovered? *Nature*, **315**, 538–539.

Diamond, J. (1986). Overview: laboratory experiments, field experiments and natural experiments. In *Community Ecology*, ed. J. Diamond and T.J. Case. New York: Harper and Row, pp. 3–22.

Dick, R.S. (1975). A map of Australia according to Koppen's principles of definition. *Queensland Geographic Journal* (3rd Ser.), **3**, 33–69.

Dickman, C.R. (1980). Ecological studies of *Antechinus stuartii* and *Antechinus flavipes* (Marsupialia: Dasyuridae) in open-forest and woodland habitats. *Australian Zoologist*, **20**, 433–446.

Dickman, C.R. and Woodford Ganf, R. (2007). *A Fragile Balance: The Extraordinary Story of Australian Marsupials*. Roseville, Sydney, Australia: Craftsman House.

Dieckmann, U., O'Hara, B. and Weisser, W. (1999). The evolutionary ecology of dispersal. *Trends in Ecology and Evolution*, **14**, 88–90.

Digby, P.G. and Kempton, R.A. (1987). *Multivariate Analysis of Ecological Communities*. London: Chapman and Hall.

Doak, D. and Mills, L.S. (1994). A useful role for theory in conservation. *Ecology*, **75**, 615–626.

Doerr, E.D. and Doerr, V.A. (2005). Dispersal range analysis: quantifying individual variation in dispersal behaviour. *Oecologia*, **142**, 1–10.

Doerr, V.A., Doerr, E.D. and Jenkins, S.H. (2006). Habitats selection in two Australasian treecreepers: what cues should they use? *Emu*, **106**, 93–103.

Drinnan, I.N. (2005). The search for fragmentation thresholds in a southern Sydney suburb. *Biological Conservation*, **124**, 339–349.

Due, A.D. and Polis, G.A. (1986). Trends in scorpion diversity along the Baja California peninsula. *American Naturalist*, **128**, 460–468.

Duncan, R.S. and Chapman, C.A. (1999). Seed dispersal and potential forest succession in abandoned agriculture in tropical Africa. *Ecological Applications*, **9**, 998–1008.

Dyck, W. (2000). Nature conservation in New Zealand plantation forestry. In *Nature Conservation 5. Nature Conservation in Production Environments: Managing the Matrix*, ed. J. Craig, N. Mitchell and D. Saunders. Chipping Norton, NSW, Australia: Surrey Beatty and Sons.

Eens, M. (1994). Bird-song as an indicator of habitat suitability. *Trends in Evolution and Ecology*, **9**, 63–64.

Ellner, S.P., Fieberg, J., Ludwig, D. and Wilcox, C. (2002). Precision of population viability analysis. *Conservation Biology*, **16**, 258–261.

Enoksson, B., Angelstam, P. and Larsson, K. (1995). Deciduous forest and resident birds: the problem of fragmentation within a coniferous forest landscape. *Landscape Ecology*, **10**, 267–275.

Epps, C.W., Paslbøll, P.J., Wehausen, J.D., Roderick, G.K., Ramey, R.R. and McCullough, D.R. (2005). Highways block gene flow and cause a rapid decline in genetic diversity in desert bighorn sheep. *Ecology Letters*, **8**, 1029–1038.

Estades, C.F. and Temple, S.A. (1999). Deciduous-forest bird communities in a fragmented landscape dominated by exotic pine plantations. *Ecological Applications*, **9**, 573–585.

Euskirchen, E.S., Chen, J.Q. and Bi, R.C. (2001). Effects of edges on plant communities in a managed landscape in northern Wisconsin. *Forest Ecology and Management*, **148**, 93–108.

Fagan, W.F., Meir, E., Prendergast, J., Folarin, A. and Karieva, P. (2001). Characterizing population vulnerability for 758 species. *Ecology Letters*, **4**, 132–138.

Fahrig, L. (1992). Relative importance of spatial and temporal scales in a patchy environment. *Theoretical Population Biology*, **41**, 300–314.

Fahrig, L. (2003). Effects of habitat fragmentation on biodiversity. *Annual Review of Ecology, Evolution and Systematics*, **34**, 487–515.

Fazey, I., Fischer, J. and Lindenmayer, D.B. (2005). Who does all the research in conservation biology? *Biodiversity and Conservation*, **14**, 917–934.

Fieberg, J. and Ellner, S.P. (2001). Stochastic matrix models for conservation and management: a comparative review of methods. *Ecology Letters*, **4**, 244–266.

Fischer, J. and Lindenmayer, D.B. (2002a). The conservation value of paddock trees for birds in a variegated landscape in southern New South Wales. I. Species composition and site occupancy patterns. *Biodiversity and Conservation*, **11**, 807–832.

Fischer, J. and Lindenmayer, D.B. (2002b). The conservation value of paddock trees for birds in a variegated landscape in southern New South Wales. II. Paddock trees as stepping stones. *Biodiversity and Conservation*, **11**, 832–849.

Fischer, J. and Lindenmayer, D.B. (2002c). The conservation value of small habitat patches: two case studies on birds from southeastern Australia. *Biological Conservation*, **106**, 129–136.

Fischer, J. and Lindenmayer, D.B. (2005a). Nestedness in fragmented landscapes: a case study on birds, arboreal marsupials and lizards. *Journal of Biogeography*, **32**, 1737–1750.

Fischer, J. and Lindenmayer, D.B. (2005b). Perfectly nested or significantly nested: an important difference for conservation management. *Oikos*, **109**, 485–494.

Fischer, J. and Lindenmayer, D.B. (2005c). The sensitivity of lizards to elevation: a case study from southeastern Australia. *Diversity and Distribution*, **11**, 225–233.

Fischer, J. and Lindenmayer, D.B. (2006). Beyond fragmentation: a new conceptual landscape model for fauna research and conservation in human-modified landscapes *Oikos*, **112**, 473–480.

Fischer, J. and Lindenmayer, D.B. (2007). Landscape modification and habitat fragmentation: a synthesis. *Global Ecology and Biogeography*, **16**, 265–280.

Fischer, J., Lindenmayer, D.B. and Fazey, I. (2004). Appreciating ecological complexity: habitat contours as a conceptual model. *Conservation Biology*, **18**, 1245–1253.

Fischer, J., Lindenmayer, D.B., Barry, S. and Flowers, E. (2005). Lizard distribution patterns in an Australian plantation landscape. *Biological Conservation*, **123**, 301–315.

Fischer, J., Lindenmayer, D.B., Blomberg, S., Montague-Frake, R. and Felton, A. (2007). Functional richness and relative resilience of bird communities in regions with different land use intensities. *Ecosystems*, **10**, 964–974.

Fischer, J., Lindenmayer, D.B. and Montague-Drake, R. (2008). The role of landscape texture in conservation biogeography: a case study on birds in south-eastern Australia. *Diversity and Distributions*, **14**, 38–46.

Fleishman, E., Betrus, C.J., Blair, R.B., Mac Nally, R. and Murphy, D.D. (2002). Nestedness analysis and conservation planning: the importance of place, environment, and life history across taxonomic groups. *Oecologia*, **133**, 78–89.

Foley, W.J. and Hume, I.D. (1987). Nitrogen requirements and urea metabolism in two arboreal marsupials, the greater glider (*Petauroides volans*) and the brushtail possum (*Trichosurus vulpecula*), fed eucalyptus foliage. *Physiological Zoology*, **60**, 241–250.

Foley, W.J., Kehl, J.C., Nagy, K.A., Kaplan I.R. and Borsoom, A.C. (1990). Energy and water metabolism in free-living greater gliders, *Petauroides volans*. *Australian Journal of Zoology*, **38**, 1–9.

Folke, C., Carpenter S., Walker, B., Scheffer, M., Elmqvist, T., Gunderson, L. and Holling, C.S. (2004). Regime shifts, resilience and biodiversity in ecosystem management. *Annual Review of Ecology Evolution and Systematics*, **35**, 557–581.

Food and Agriculture Organization of the United Nations (2001). *State of the World's Forests*. Rome, Italy: Food and Agriculture Organization of the United Nations.

Food and Agriculture Organization of the United Nations (2007). *State of the World's Forests*. Rome, Italy: Food and Agriculture Organization of the United Nations.

Ford, H.A., Barrett, G.W., Saunders. D.A. and Recher, H.F. (2001). Why have birds in the woodlands of southern Australia declined? *Conservation Biology*, **97**, 71–88.

Forman, R.T. (1995). *Land Mosaics: The Ecology of Landscapes and Regions*. New York: Cambridge University Press.

Forman, R.T. and Godron, M. (1986). *Landscape Ecology*. New York: Wiley and Sons.

Forman, R.T., Sperling, D., Bissonette, J.A., Clevenger, A.P., Cutshall, C.D., Dale, V.H., Fahrig, L., France, R., Goldman, C.R., Heanue, K., Jones, J.A., Swanson, F.J., Turrentine, T. and Winter, T.C. (eds.) (2002). *Road Ecology: Science and Solutions*. Washington, DC: Island Press.

Fox, B.J. and McKay, G.M. (1981). Small mammal responses to pyric successional changes in eucalypt forests. *Australian Journal of Ecology*, **6**, 29–41.

Franklin, J.F. (2003). Challenges to temperate forest stewardship: focusing on the future. In *Towards Forest Sustainability*, ed. D.B. Lindenmayer and J.F. Franklin. Washington, DC: Island Press, pp. 1–13.

Franklin, J.F. and MacMahon, J.A. (2000). Messages from a mountain. *Science*, **288**, 1183–1185.

Franklin, J.F., Lindenmayer, D.B., MacMahon, J.A., McKee, A., Magnusson, J., Perry, D.A., Waide, R. and Foster, D.R. (2000). Threads of continuity: ecosystem disturbances, biological legacies and ecosystem recovery. *Conservation Biology in Practice*, **1**, 8–16.

Friend, G.R. (1980). Wildlife conservation and softwood forestry in Australia: some considerations. *Australian Forestry*, **43**, 217–224.

Friend, G.R. (1982a). Mammal populations in exotic pine plantations and indigenous eucalypt forests in Gippsland, Victoria. *Australian Forestry*, **45**, 3–18.

Friend, G.R. (1982b). Bird populations in exotic pine plantations and indigenous eucalypt forests in Gippsland, Victoria. *Emu*, **45**, 80–91.

Gall, B.C. (1982). *Wildlife in the South-western Slopes Region of NSW*. Report to NSW National Parks and Wildlife Service. Sydney, Australia: NSW National Parks and Wildlife Service.

Gascon, C., Lovejoy, T.E., Bierregaard, R.O.J., Malcolm, J.R., Stouffer, P.C., Vasconcelos, H.L., Laurance, W.F., Zimmerman, B., Tocher, M. and Borges, S. (1999). Matrix habitat and species richness in tropical forest remnants. *Biological Conservation*, **91**, 223–229.

Gaston, K.J. and Spicer, J.I. (2004). *Biodiversity: An Introduction*, 2nd edn. Oxford, UK: Blackwell Publishing.

Gates, J.E. and Gysel, L.W. (1978). Avian nest dispersion and fledging success in field-forest ecotones. *Ecology*, **59**, 871–883.

Gibbons, P. and Lindenmayer, D.B. (2002). *Tree Hollows and Wildlife Conservation in Australia*. Melbourne, Australia: CSIRO Publishing.

Gibbons, P. and Lindenmayer, D.B. (2007). The use of offsets to regulate land clearing: no net loss or the tail wagging the dog? *Environmental Management and Restoration*, **8**, 26–31.

Gilmore, A.M. (1985). The influence of vegetation structure on the density of insectivorous birds. In *Birds of Eucalypt Forests and Woodlands: Ecology, Conservation, Management*, ed. A. Keast, H.F. Recher, H. Ford and D. Saunders. Chipping Norton, NSW, Australia: Surrey Beatty and Sons, pp. 21–31.

Gilpin, M.E. and Diamond, J.M. (1980). Subdivision of nature reserves and the maintenance of species diversity. *Nature*, **285**, 567–568.

Gilpin, M.E. and Soulé, M.E. (1986). Minimum viable populations: processes of species extinctions. In: *Conservation Biology: The Science of Scarcity and Diversity*, ed. M.E. Soulé. Sunderland, MA: Sinauer Associates, pp. 19–34.

Golden, D.M. and Crist, T.O. (1999). Experimental effects of habitat fragmentation on old-field canopy insects: community, guild and species responses. *Oecologia*, **118**, 371–380.

Goldingay, R.L. (1986). Feeding behaviour of the yellow-bellied glider, *Petaurus australis* (Marsupialia: Petauridae) at Bombala, New South Wales. *Australian Mammalogy*, **9**, 17–25.

Goldingay, R.G. and Kavanagh, R.P. (1991). The yellow-bellied glider: a review of its ecology and management considerations. In *Conservation of Australia's Forest Fauna*, ed. D. Lunney. Sydney, Australia: Surrey Beatty and Sons, pp. 365–375.

Goldingay, R.G. and Kavanagh, R.P. (1993). Home range estimates of the yellow-bellied glider (*Petaurus australis*) at Waratah Creek, New South Wales. *Wildlife Research*, **20**, 387–404.

Graves, G.R., and Gotelli, N.J. (1993). Assembly of avian mixed-species flocks in Amazonia. *Proceedings of the National Academy of Sciences of the USA*, **90**, 1388–1391.

Green, K. and Osborne, W. (1994). *Wildlife of the Australian Snow-country*. Sydney, Australia: Reed Books.

Groffman, P.M., Baron, J.S., Blett, T., Gold, A.J., Goodman, I., Gunderson, L.H., Levinson, B.M., Palmer, M.A., Paerl, H.W., Peterson, G.D., Poff, N.L., Regeski, D.W., Reynolds, J.F., Turner, M.G., Weathers, K.C. and Wiens, J. (2006). Ecological thresholds: the key to successful environmental management or an important concept with no practical application? *Ecosystems*, **9**, 1–13.

Groom, M., Meffee, G.K., Carroll, C.R. (2005). *Principles of Conservation Biology*, 3rd edn. Sunderland, MA: Sinauer Associates.

Gustafson, E.J. (1998). Quantifying landscape spatial pattern: what is the state of the art? *Ecosystems*, **1**, 143–156.

Haila, Y. (2002). A conceptual genealogy of fragmentation research from island biogeography to landscape ecology. *Ecological Applications*, **12**, 321–334.

Haines-Young, R. and Chopping, M. (1996). Quantifying landscape structure: a review of landscape indices and their application to forested environments. *Progress in Physical Geography*, **20**, 418–445.

Hall, L.S., Krausman, P.A. and Morrison, M.L. (1997). The habitat concept and a plea for the use of standard terminology. *Wildlife Society Bulletin*, **25**, 173–182.

Hall, S. (1980). The diets of two co-existing species of *Antechinus* (Marsupialia: Dasyuridae). *Australian Wildlife Research*, **7**, 365–378.

Hannon, S.J., and Schmiegelow, F. (2002). Corridors may not improve the conservation value of small reserves for most boreal birds. *Ecological Applications*, **12**, 1457–1468.

Hanski, I. (1994a). Patch occupancy dynamics in fragmented landscapes. *Trends in Ecology and Evolution*, **9**, 131–134.

Hanski, I. (1994b). A practical model of metapopulation dynamics. *Journal of Animal Ecology*, **63**, 151–162.

Hanski, I. (1997). Metapopulation dynamics: from concepts and observations to predictive models. In *Metapopulation Biology: Ecology, Genetics and Evolution*. ed. I. Hanski and M.E. Gilpin. San Diego, CA: Academic Press, pp. 69–91.

Hanski, I. (1999). *Metapopulation Ecology*. Oxford, UK: Oxford University Press.

Hanski, I. and Gilpin, M. (1991). Metapopulation dynamics: brief history and conceptual domain. *Biological Journal of the Linnean Society*, **42**, 3–16.

Hanski, I. and Simberloff, D. (1997). The metapopulation approach: its history, conceptual domain and application to conservation. In *Metapopulation Biology: Ecology, Genetic and Evolution*, ed. I. Hanski and M.E. Gilpin. San Diego, CA: Academic Press, pp. 5–26.

Hansson, L. (1998). Vertebrate distributions relative to clear-cut edges in a boreal forest landscape. *Landscape Ecology*, **9**, 105–115.

Harper, K.A., Macdonald, S.E., Burton, P.J., Chen, J., Brosofske, K.D., Saunders, S.C., Euskirchen, E.S., Roberts, D., Jaiteh, M.S. and Essen, P-E. (2005). Edge influence on forest structure and composition in fragmented landscapes. *Conservation Biology*, **19**, 768–782.

Harrison, S. (1991). Local extinction in a metapopulation context: an empirical evaluation. *Biological Journal of the Linnean Society*, **42**, 73–88.

Haskell, D.G., Evans, J.P. and Pelkey, N.W. (2006). Depauperate avifauna in plantations compared to forests and exurban areas. *PLoS ONE*, **1**, e63.

Hastie, T.J. and Tibshirani, R.J. (1990). *Generalized Additive Models*. London: Chapman and Hall.

Haynes, R.W., Bormann, B.T., Lee, D.C. and Martin, J.R. (2006). *Northwest Forest Plan – The First 10 Years (1993–2003): Synthesis of Monitoring and Research Results*. General Technical Report PNW-GTR-651, October, 2006. Pacific Northwest Research Station, Portland OR: USDA Forest Service.

Henry, S.R, and Craig, S.A. (1984). Diet, ranging behaviour and social organisation of the yellow-bellied glider (*Petaurus australis* Shaw) in Victoria. In *Possums and Gliders*, ed. A.P. Smith and I.D. Hume. Sydney, Australia: Surrey Beatty and Sons for the Australian Mammal Society, pp. 331–341.

Higgins, P. (ed.) (1999). *Handbook of Australian New Zealand and Antarctic Birds. Volume 4: Parrots to Dollarbird*. Melbourne, Australia: Oxford University Press.

Hilty, J., Lidicker, W., Jr and Merenlender, A. (2006). *Corridor Ecology: The Science and Practice of Linking Landscapes for Biodiversity Conservation*. Washington, DC: Island Press.

Hobbs, R.J. and Yates, C.J. (2003). Impacts of ecosystem fragmentation on plant populations: generalising the idiosyncratic. *Australian Journal of Botany*, **51**, 471–488.

Holling, C.S. (1992). Cross-scale morphology, geometry, and dynamics of ecosystems. *Ecological Monographs*, **62**, 447–502.

Homan, R.N., Windmiller, B.S. and Reed, J.M. (2004). Critical thresholds associated with habitat loss for two vernal pool-breeding amphibians. *Ecological Applications*, **14**, 1547–1553.

Honnay, O., Hermy, M. and Coppin, P. (1999). Effects of area, age and diversity of forest patches in Belgium on plant species richness, and implications for conservation and reafforestation. *Biological Conservation*, **87**, 73–84.

Honnay, O., Verheyen, K. and Hermy, M. (2002). Permeability of ancient forest edges for weedy plant species invasion. *Forest Ecology and Management*, **161**, 109–122.

Hopper, S.D. (2000). Floristics of Australian granitoid inselberg vegetation. In *Inselbergs*, ed. S. Porembski and W. Barthlott. Ecological Studies Series, Volume 146. Berlin, Germany: Springer, pp. 391–407.

How, R. (1972). The ecology and management of *Trichosurus* spp. (Marsupialia) in NSW. Unpublished Ph.D. thesis, University of New England, Armidale, NSW, Australia.

How, R.A. (1981). Population parameters of two congeneric possums, *Trichosurus* spp., in north-eastern New South Wales. *Australian Journal of Zoology*, **29**, 205–15.

How, R.A., Barnett, J.L., Bradley, A.J., Humphreys, W.F. and Martin, R. (1984). The population biology of *Pseudocheirus peregrinus* in a *Leptospermum laevigatum* thicket. In *Possums and Gliders*, ed. A.P. Smith and I.D. Hume. Sydney, Australia: Surrey Beatty and Sons, pp. 261–268.

Howe, R.W. (1984). Local dynamics of bird assemblages in small forest habitat islands in Australia and North America. *Ecology*, **65**, 1585–1601.

Huggett, A.J. (2005). The concept and utility of 'ecological thresholds' in biodiversity conservation. *Biological Conservation*, **124**, 301–310.

Huhta, E., Jokimäki, J. and Rahko, P. (1999). Breeding success of pied flycatchers in artificial forest edges: the effect of a suboptimally shaped foraging area. *Auk*, **116**, 528–535.

Jackson, R.B. and Schlesinger, W.H. (2004). Curbing the U.S. carbon deficit. *Proceedings of the National Academy of Sciences*, **101**, 15827–15829.

Jackson, R.B., Jobbagy, E.G., Avissar, R., Roy, S.B., Barrett, D.J., Cook, C.W., Farley, K.A., le Maitre, D.C., McCarl, B.A. and Murray, B.C. (2005). Trading water for carbon with biological sequestration. *Science*, **310**, 1944–1947.

Jacobs, M.R. (1955). *Growth Habits of the Eucalypts*. Forestry & Timber Bureau, Department of the Interior, Canberra, Australia.

Janzen, D.H. (1983). No park is an island: increase in interference from outside as park size decreases. *Oikos*, **41**, 402–410.

Jenkins, R. and Bartell, R. (1980). *A Field Guide to the Reptiles of the Australian High Country*. Melbourne, Australia: Inkata Press.

Joyal, L., McCollough, A.M. and Hunter, M.L. (2001). Landscape ecology approaches to wetland species conservation: a case study of two turtle species in southern Maine. *Conservation Biology*, **15**, 1755–1762.

Kavanagh, R.P. and Lambert, M.J. (1990). Food selection by the greater glider, *Petauroides volans*: is foliar nitrogen a determinant of habitat quality? *Australian Wildlife Research*, **17**, 285–299.

Kehl, J.A. and Borsboom, A. (1984). Home ranges, den use and activity patterns in the greater glider *Petauroides volans*. In *Possums and Gliders*, ed. A.P. Smith and I.D. Hume. Sydney, Australia: Surrey Beatty and Sons, pp. 229–236.

Keogh, J.S., Webb, J.K. and Shine, R. (2007). Spatial genetic analysis and long-term mark–recapture data demonstrate male-biased dispersal in a snake. *Biology Letters*, **3**, 33–35.

Kerle, A. (2001). *Possums: The Brushtails, Ringtails and Greater Glider*. Sydney, Australia: UNSW Press.

Kiester, A.R. (1971). Species density of North American amphibians and reptiles. *Systematic Zoology*, **20**, 127–137.

Koenig, W.D. (1998). Spatial autocorrelation in California land birds. *Conservation Biology*, **12**, 612–619.

Kotliar, N.B. and Wiens, J.A. (1990). Multiple scales of patchiness and patch structure: a hierarchical framework for the study of heterogeneity. *Oikos*, **59**, 253–260.

Kraaijeveld-Smit, F.J.L., Lindenmayer, D.B., Taylor, A.C., MacGregor, C. and Wertheim, B. (2007). Genetic structure and dispersal patterns of three sympatric small mammal species. *Oikos*, **116**, 1819–1830.

Krebs, C.J. (2008). *The Ecological World View*. Melbourne, Australia: CSIRO Publishing.

Kremsater, L. and Bunnell, F.L. (1999). Edge effects: theory, evidence and implications to management of western North American forests. In *Forest Wildlife and Fragmentation: Management Implication*, ed. J. Rochelle, L.A. Lehmann and J. Wisniewski. Leiden, Germany: Brill, pp. 117–153.

Kreuzer, M.P. and Huntly, N.J. (2003). Habitat-specific demography: evidence for source–sink population structure in a mammal, the pika. *Oecologia*, **134**, 343–349.

Krohne, D.T. (1997). Dynamics of metapopulations of small mammals. *Journal of Mammalogy*, **78**, 1014–1026.

Lacy, R.C. (2000). Structure of the VORTEX simulation model for population viability analysis. *Ecological Bulletins*, **48**, 191–203.

Lacy, R.C., Flesness, N.R. and Seal, U.S. (1989). *Puerto Rican Parrot Population Viability Analysis*. Report to the US Fish and Wildlife Service. Apple Valley, MN: Captive Breeding Specialist Group, Species Survival Commission, IUCN.

Lambeck, R.J. (1997). Focal species: a multi-species umbrella for nature conservation. *Conservation Biology*, **11**, 849–856.

Lambeck, R.J. (1999). *Landscape Planning for Biodiversity Conservation in Agricultural Regions: A Case Study from the Wheatbelt of Western Australia*. Biodiversity Technical Paper No. 2. Canberra, Australia: Environment Australia.

Lamberson, R.H., Noon, B.R., Voss, C. and McKelvey, R. (1994). Reserve design for territorial species: the effects of patch size and spacing on the viability of the Northern Spotted Owl. *Conservation Biology*, **8**, 185–195.

Lamm, D.W. and Calaby, J.H. (1950). Seasonal variation of bird populations along the Murrumbidgee in the Australian Capital Territory. *Emu*, **50**, 114–122.

Landres, P.B., Verner, J. and Thomas, J.W. (1988). Ecological uses of vertebrate indicator species: a critique. *Conservation Biology*, **2**, 316–328.

Landsberg, J. (1999). Status of temperate eucalypt woodlands in the Australian Capital Territory region. In *Temperate Eucalypt Woodlands in Australia: Biology, Conservation, Management and Restoration*, ed. R.J. Hobbs and C.J. Yates. Chipping Norton, NSW, Australia: Surrey Beatty and Sons, pp. 32–44.

Laurance, W.F. (1990). Comparative responses of five arboreal marsupials to forest fragmentation. *Journal of Mammalogy*, **71**, 641–653.

Laurance, W.F. (1991). Edge effects in tropical forest fragments: application of a model for the design of nature reserves. *Biological Conservation*, **57**, 205–219.

Laurance, W.F. (2000). Do edge effects occur over large spatial scales? *Trends in Ecology and Evolution*, **15**, 134–135.

Laurance, W.F., Bierregaard, R.O., Gascon, C., Didham, R.K., Smith, A.P., Lynam, A.J., Viana, V.M., Lovejoy, T.E., Sieving, K.E., Sites, J.W., Andersen, M., Tocher, M.D., Kramer, E.A., Restrepo, C. and Moritz, C. (1997). Tropical forest fragmentation: synthesis of a diverse and dynamic discipline. In *Tropical Forest Remnants: Ecology, Management and Conservation of Fragmented Communities*, ed. W.F. Laurance and R.O. Bierregaard. Chicago, IL: The University of Chicago Press, pp. 502–525.

Laurance, W.F., Perez-Salicrup, D., Delamonica, P., Fearnside, P.M., D'Angelo, S., Jerozolinski, A., Pohl, L. and Lovejoy, T.E. (2001). Rain forest fragmentation and the structure of Amazonian Liana communities. *Ecology*, **82**, 105–116.

Law, B. and Dickman, C.R. (1998). The use of habitat mosaics by terrestrial vertebrate fauna: implications for conservation and management. *Biodiversity and Conservation*, **7**, 323–333.

Lawlor, T.E. (1983). The peninsula effect on mammalian species diversity in Baja California. *American Naturalist* **121**, 432–439.

Lazenby-Cohen, K.A. (1991). Communal nesting in *Antechinus stuartii* (Marsupialia: Dasyuridae). *Australian Journal of Zoology*, **39**, 273–283.

Lee, A.K. and Cockburn, A. (1985). *Evolutionary Ecology of Marsupials*. Sydney, Australia: Cambridge University Press.

Legge, S and Cockburn, A. (2000). Social and mating system of cooperatively breeding laughing kookaburras (*Dacelo novaeguineae*). *Behavioral Ecology and Sociobiology*, **47**, 220–229.

Lemckert, F. (1998). A survey of the threatened herpetofauna of the south-west slopes of New South Wales. *Australian Zoologist*, **30**, 492–500.

Lenoir, L., Bengtsson, J. and Persson, T. (2003). Effects of *Formica* ants on soil fauna: results from a short-term exclusion and a long-term natural experiment. *Oecologia*, **134**, 423–430.

Leung, K.P., Dickman, C.R. and Moore, L.A. (1993). Genetic variation in fragmented populations of an Australian rainforest rodent, *Melomys cervinipes*. *Pacific Conservation Biology*, **1**, 58–65.

Levey, D.J., Bolker, B.M., Tewksbury, J.J., Sargent, S. and Haddad, N.M. (2005). Effects of landscape corridors on seed dispersal by birds. *Science*, **309**, 146–148.

Lindenmayer, D.B. (1997). Differences in the biology and ecology of arboreal marsupials in forests of south-eastern Australia. *Journal of Mammalogy*, **78**, 1117–1127.

Lindenmayer, D.B. (2000). *Islands of Bush in a Sea of Pines: The Tumut Fragmentation Experiment to August 2000*. Research Report 6/00. Canberra, Australia: Land and Water Resources Research and Development Corporation.

Lindenmayer, D.B. (2002). *Gliders of Australia: A Natural History*. Sydney, Australia: UNSW Press.

Lindenmayer D.B. and Burgman, M.A. (2005). *Practical Conservation Biology*. Melbourne, Australia: CSIRO Publishing.

Lindenmayer D.B and Fischer, J. (2006). *Landscape Change and Habitat Fragmentation*. Washington, DC: Island Press.

Lindenmayer, D.B. and Fischer, J. (2007). Tackling the habitat fragmentation panchreston. *Trends in Ecology and Evolution*, **22**, 127–132.

Lindenmayer, D.B. and Franklin, J.F. (2002). *Conserving Forest Biodiversity: A Comprehensive Multiscaled Approach*. Washington, DC: Island Press.

Lindenmayer, D.B. and Hobbs, R.J. (2004). Biodiversity conservation in plantation forests: a review with special reference to Australia. *Conservation Biology*, **119**, 151–168.

Lindenmayer, D.B. and Hobbs, R.J. (eds.) (2007). *Managing and Designing Landscapes for Conservation: Moving from Perspectives to Principles*. Oxford, UK: Blackwell Publishing.

Lindenmayer, D.B. and Lacy, R.C. (2002). Small mammals, patches and PVA models: a field test of model predictive ability. *Biological Conservation*, **103**, 247–265.

Lindenmayer, D.B. and McCarthy, M.A. (2001). The spatial distribution of non-native plant invaders in a pine–eucalypt mosaic in south-eastern Australia. *Biological Conservation*, **102**, 77–87.

Lindenmayer, D.B. and Peakall, R.H. (2000). The Tumut experiment: integrating demographic and genetic studies to unravel fragmentation effects. In *Genetics, Demography and Viability of Fragmented Populations*, ed. A. Young and G. Clarke. Cambridge, UK: Cambridge University Press, pp. 173–201.

Lindenmayer, D.B., Cunningham, R.B., Tanton, M.T., Smith, A.P. and Nix, H.A. (1990). The habitat requirements of the mountain brushtail possum and the greater glider in the montane ash-type eucalypt forests of the Central Highlands of Victoria. *Australian Wildlife Research*, **17**, 467–478.

Lindenmayer, D.B., Cunningham, R.B. and Donnelly, C.F. (1993). The conservation of arboreal marsupials in the montane ash forests of the Central Highlands of Victoria, south-east Australia. IV. The distribution and abundance of arboreal marsupials in retained linear strips (wildlife corridors) in timber production forests. *Biological Conservation*, **66**, 207–221.

Lindenmayer, D.B., Ritman, K.R., Cunningham, R.B., Smith, J. and Howarth, D. (1995a). A method for predicting the spatial distribution of arboreal marsupials. *Wildlife Research*, **22**, 445–456.

Lindenmayer, D.B., Burgman, M.A., Açkakaya, H.R., Lacy, R.C. and Possingham, H.P. (1995b). A review of the generic computer programs ALEX, RAMAS/Space and VORTEX for modelling the viability of wildlife metapopulations. *Ecological Modelling*, **82**, 161–174.

Lindenmayer, D.B., Pope, M., Cunningham, R.B., Donnelly, C.F., and Nix, H.A. (1996). Roosting of the Sulphur-Crested Cockatoo *Cacatua galerita*. *Emu*, **96**, 209–212.

Lindenmayer, D.B., Cunningham, R.B., Pope, M., Donnelly C.F., Nix, H.A. and Incoll R.D. (1997a). *The Response of Arboreal Marsupials to Context: A Large-Scale Fragmentation Study at Tumut, Southeastern Australia*. Working Paper 1997/4. Canberra, Australia: Centre for Resource and Environmental Studies, The Australian National University.

Lindenmayer, D.B., Cunningham, R.B., Nix, H.A., Lindenmayer, B.D., McKenzie, S., McGregor, C., Pope, M.L. and Incoll, R.D. (1997b). *Counting Birds in Forests: A Comparison of Observers and Observation Methods*. CRES Working Paper. Canberra, Australia: Centre for Resource and Environmental Studies, The Australian National University.

Lindenmayer, D.B., Cunningham, R.B., Pope, M. and Donnelly, C.F. (1999a). The response of arboreal marsupials to landscape context: a large-scale fragmentation study. *Ecological Applications*, **9**, 594–611.

Lindenmayer, D.B., Pope, M.L. and Cunningham, R.B. (1999b). Roads and nest predation: an experimental study in a modified forest ecosystem. *Emu*, **99**, 148–152.

Lindenmayer, D.B., Cunningham, R.B., Pope, M.L. and Donnelly, C.F. (1999c). A large-scale 'experiment' to examine the effects of landscape context and habitat fragmentation on mammals. *Biological Conservation*, **88**, 387–403.

Lindenmayer, D.B., Cunningham, R.B., Incoll, R.D., Pope, M.L., Donnelly, C.F., MacGregor, C., Tribolet, C.R. and Triggs, B.E. (1999d). Comparison of hairtube types for the detection of mammals. *Wildlife Research*, **26**, 745–753.

Lindenmayer, D.B., McCarthy, M.A. and Pope, M.L. (1999e). Arboreal marsupial incidence in eucalypt patches in south-eastern Australia: a test of Hanski's incidence function metapopulation model for patch occupancy. *Oikos*, **84**, 99–109.

Lindenmayer, D.B., Lacy, R.C., Tyndale-Biscoe, H., Taylor, A., Viggers, K. and Pope, M.L. (1999f). Integrating demographic and genetic studies of the greater glider (*Petauroides volans*) at Tumut, south-eastern Australia: setting hypotheses for future testing. *Pacific Conservation Biology*, **5**, 2–8.

Lindenmayer, D.B., Margules, C.R. and Botkin, D. (2000a). Indicators of forest sustainability biodiversity: the selection of forest indicator species. *Conservation Biology*, **14**, 941–950.

Lindenmayer, D.B., Lacy, R.C. and Pope, M.L. (2000b). Testing a simulation model for population viability analysis. *Ecological Applications*, **10**, 580–597.

Lindenmayer, D.B., Cunningham, R.B., Pope, M.L., Gibbons, P. and Donnelly, C. F. (2000c). Cavity sizes and types in Australian wet and dry eucalypt forest: a simple rule of thumb. *Forest Ecology and Management*, **137**, 139–150.

Lindenmayer, D.B., Cunningham, R.B., Donnelly, C.F., Incoll, R.D., Pope, M.L., Tribolet, C.R., Viggers, K.L. and Welsh, A.H. (2001a). How effective is spotlighting for detecting the greater glider (*Petauroides volans*)? *Wildlife Research*, **28**, 105–109.

Lindenmayer, D.B., Ball, I., Possingham, H.P., McCarthy, M.A. and Pope, M.L. (2001b). A landscape-scale test of the predictive ability of a meta-population model in an Australian fragmented forest ecosystem. *Journal of Applied Ecology*, **38**, 36–48.

Lindenmayer, D.B., McCarthy, M.A., Possingham, H.P. and Legge, S. (2001c). A simple landscape-scale test of a spatially explicit population mode: patch occupancy in fragmented south-eastern Australian forests. *Oikos*, **92**, 445–458.

Lindenmayer, D.B., Cunningham, R.B., Donnelly, C.F. and Nix, H.A. (2002a). The distribution of birds in a novel landscape context. *Ecological Monographs*, **72**, 1–18.

Lindenmayer, D.B., Cunningham, R.B., Donnelly, C.F. and Lesslie, R. (2002b). On the use of landscape indices as ecological indicators in fragmented forests. *Forest Ecology and Management*, **159**, 203–216.

Lindenmayer, D.B., Dubach, J. and Viggers, K.L. (2002c) Geographic dimorphism in the mountain brushtail possum (*Trichosurus caninus*): the case for a new species. *Australian Journal of Zoology*, **50**, 369–393.

Lindenmayer, D.B., Possingham, H.P., Lacy, R.C., McCarthy, M.A. and Pope, M. L. (2003a). How accurate are population models? Lessons from landscape-scale population tests in a fragmented system. *Ecology Letters*, **6**, 41–47.

Lindenmayer, D.B., McIntyre, S. and Fischer, J. (2003b). Birds in eucalypt and pine forests: landscape alteration and its implications for research models of faunal habitat use. *Biological Conservation*, **110**, 45–53.

Lindenmayer, D.B., Cunningham, R.B. and Lindenmayer, B.D. (2004a). Sound recording of bird vocalisations in forests. II. Longitudinal profiles in vocal activity. *Wildlife Research*, **31**, 209–217.

Lindenmayer, D.B., Pope, M.L., and Cunningham, R.B. (2004b). Patch use by the greater glider (*Petauroides volans*) in a fragmented forest ecosystem. II. Characteristics of den trees and preliminary data on den-use patterns. *Wildlife Research*, **31**, 569–577.

Lindenmayer, D.B., Cunningham, R.B. and Peakall, R. (2005a). On the recovery of populations of small mammals in forest fragments following major population reduction. *Journal of Applied Ecology*, **42**, 649–658.

Lindenmayer, D.B., Cunningham, R.B. and Fischer, J. (2005b). Native vegetation cover thresholds with species responses. *Biological Conservation*, **124**, 311–316.

Lindenmayer D.B., Franklin, J.F. and Fischer, J. (2006). Conserving forest biodiversity: a checklist for forest and wildlife managers. *Biological Conservation*, **129**, 511–518.

Lindenmayer, D.B., Hobbs, R.J., Montague-Drake, R., Alexandra, J., Bennett, A., Burgman, M., Cale, P., Calhoun, A., Cramer, V., Cullen, P., Driscoll, D., Fahrig, L., Fischer, J., Franklin, J., Gibbons, P., Haila, Y., Hunter, M., Lake, S., Luck, G., McIntyre, S., Mac Nally, R., Manning, A., Miller, J., Mooney, H., Noss, R., Possingham, H., Saunders, D., Schmiegelow, F., Scott, M., Simberloff, D., Sisk, T., Walker, B., Wiens, J., Woinarski, J. and Zavaleta, E. (2007a). A checklist for ecological management of landscapes for conservation. *Ecology Letters*, **11**, 78–91.

Lindenmayer, D.B., Blackmore, C., Blomberg, S., Elliott, C.P., Fischer, J., Felton, A., Felton, A.M., Lowe, A., Manning, A., Montague-Drake, R., Saunders, D., Simberloff, D., Wilson, D. and Youngentob, K. (2007b). The complementarity of single-species and ecosystem-oriented research in conservation research. *Oikos*, **116**, 1220–1226.

Lindenmayer, D.B., Cunningham, R.B., Crane, M., Lindenmayer, B.D., MacGregor, C., Michael, D. and Montague-Drake, R. (2007c). Birds surveyed in eucalypt forests, pine plantation forests and eucalypt remnants surrounded by pine stands at Tumut, south-eastern Australia. *Check List*, **3**, 168–174.

Lindenmayer, D.B., Cunningham, R.B. and Weekes, A. (2007d). Foraging ecology of the White-throated Treecreeper (*Cormobates leucophaeus*). *Emu*, **107**, 135–142.

Lindenmayer, D.B., Wood, J.T., Cunningham, R.B., Crane, M., Macgregor, C., Michael, D. and Montague-Drake, R. (2008a). Experimental evidence of the effects of a changed matrix on conserving biodiversity within patches of native forest in an industrial plantation landscape. *Landscape Ecology*, **23**, doi:10.1007/s10980-008-9244-5.

Lindenmayer, D.B., Cunningham, R.B., Crane, M., McGregor, C. and Michael, D. (2008b). The changing nature of bird populations in woodland remnants as a pine plantation emerges: results from a large-scale 'natural experiment' of landscape context effects. *Ecological Monographs*, **78**, in press.

Lindenmayer, D.B., Burton, P. and Franklin, J.F. (2008c). *Salvage Logging and its Ecological Consequences*. Washington, DC: Island Press.

Lovejoy, T.E., Bierregaard, R.O.J., Brown, K.S.J., Harper, L.H., Hays, M.B., Malcolm, J.R., Powell, A.H., Powell, G.V.N., Quintela, C.E., Rylands, A.B. and Schubart, H.O.R. (1986). Edge and other effects of isolation on Amazon forest fragments. In *Conservation Biology: The Science of Scarcity and Diversity*, ed. M.E. Soule. Sunderland, MA: Sinauer Associates, pp. 258–285.

Loyn, R.H. (1987). Effects of patch area and habitat on bird abundances, species numbers and tree health in fragmented Victorian forests. In *Nature Conservation: The Role of Remnants of Native Vegetation*, ed. G. Arnold, A. Burbidge, A. Hopkins and D.A. Saunders. Chipping Norton, NSW, Australia: Surrey Beatty and Sons, pp. 65–77.

Luck, G.W., Paton, D.C. and Possingham, H.P. (1999). Bird responses at inherent and induced edges in the Murray Mallee, South Australia. 1. Differences in abundance and diversity. *Emu*, **99**, 157–169.

Ludwig, D. (1999). Is it meaningful to estimate a probability of extinction? *Ecology*, **80**, 298–310.

Ludwig, J., Tongway, D., Freudenberger, D., Noble, J. and Hodgkinson, K.C. (1997). *Landscape Ecology, Function and Management: Principles from Australia's Rangelands*. Melbourne, Australia: CSIRO Publishing.

270 · References

Lumsden, L. F. and Bennett, A. F. (2005). Scattered trees in rural landscapes: foraging habitat for insectivorous bats in south-eastern Australia. *Biological Conservation*, **122**, 205–222.

Lumsden, L.F., Bennett, A.F., Krasna, S. and Silins, J. (1994). *Fauna in a Remnant Vegetation–Farmland Mosaic: Movement, Roosts and Foraging Ecology of Bats*. Report to the Australian Nature Conservation Agency 'Save the Bush Program'. Melbourne, Australia: Flora and Fauna Branch, Department of Conservation and Natural Resources.

Lunney, D. (1983). Bush rat. In *The Complete Book of Australian Mammals*, ed. R. Strahan. Sydney, Australia: Angus and Robertson, pp. 443–445.

MacArthur, R. H. and MacArthur, J. W. (1961). On bird species diversity. *Ecology*, **42**, 594–598.

MacArthur, R. H. and Wilson, E. O. (1963). An equilibrium theory of insular zoogeography. *Evolution*, **17**, 373–387.

MacArthur, R. H. and Wilson, E. O. (1967). *The Theory of Island Biogeography*. Princeton, NJ: Princeton University Press.

Mac Nally, R. (1994). Habitat-specific guild structure of forest birds in south-eastern Australia: a regional scale perspective. *Journal of Animal Ecology*, **63**, 988–1001.

Mac Nally, R., Bennett, A. F. and Horrocks, G. (2000). Forecasting the impacts of habitat fragmentation: evaluation of species-specific predictions of the impact of habitat fragmentation on birds in the box–ironbark forests of central Victoria, Australia. *Biological Conservation*, **95**, 7–29.

Maes, D. and Van Dyck, H. (2005). Habitat quality and biodiversity indicator performances of a threatened butterfly versus a multispecies group for wet heathlands in Belgium. *Biological Conservation*, **123**, 177–187.

Magara, T. (2002). Carabids and forest edge: spatial pattern and edge effect. *Forest Ecology and Management*, **157**, 23–37.

Major, R. E. and Kendal, C. E. (1996). The contribution of artificial nest experiments to understanding avian reproductive success: a review of methods and conclusions. *Ibis*, **138**, 298–307.

Manel, S., Schwartz, M. K., Luikart, G. and Taberlet, P. (2003). Landscape genetics: combining landscape ecology and population. *Trends in Ecology and Evolution*, **15**, 209–295.

Manning, A. D., Lindenmayer, D. B. and Nix, H. A. (2004). Continua and umwelt: alternative ways of viewing landscapes. *Oikos*, **104**, 621–628.

Manning, A. D., Fischer, J. F. and Lindenmayer, D. B. (2006). Scattered trees are keystone structures. *Biological Conservation*, **206**, 311–321.

Mansergh, I. M. and Scotts, D. J. (1989). Habitat continuity and social organization of the mountain pygmy-possum restored by tunnel. *Journal of Wildlife Management*, **53**, 701–707.

Margules, C. R. (1992). The Wog Wog habitat fragmentation experiment. *Environmental Conservation*, **19**, 316–325.

Marples, T. G. (1973). Studies of the marsupial glider, *Schoinobates volans* (Kerr). IV. Feeding biology. *Australian Journal of Zoology*, **21**, 213–216.

Martínez-Sánchez, J. J., Ferrandis, P., de las Heras, J. and Herranz, J. M. (1999). Effect of burnt wood removal on the natural regeneration of *Pinus halepensis* after fire in a pine forest in Tus Valley (SE Spain). *Forest Ecology and Management*, **123**, 1–10.

May, R.M. (1978). Evolution of ecological systems. *Scientific American*, **239**, 160.

McAlpine, C.M., Fensham, R.J. and Temple-Smith, D.E. (2002a). Biodiversity conservation and vegetation clearing in Queensland: principles and thresholds. *Rangelands Journal*, **24**, 36–55.

McAlpine, C.M., Lindenmayer, D.B., Eyre, T. and Phinn, S. (2002b). Landscape surrogates for conserving Australia's forest fauna: synthesis of Montreal Process case studies. *Pacific Conservation Biology*, **8**, 108–120.

McCarthy, M.A. and Lindenmayer, D.B. (1999). Conservation of the greater glider (*Petauroides volans*) in forests managed for timber production: implications for plantation design and remnant native vegetation retention. *Animal Conservation*, **2**, 203–209.

McCarthy, M.A., Lindenmayer, D.B. and Possingham, H.P. (2000). Testing spatial PVA models of Australian treecreepers (Aves: Climacteridae) in fragmented forest. *Ecological Applications*, **10**, 1722–1731.

McCarthy, M.A., Lindenmayer, D.B., and Possingham, H.P. (2001). Assessing spatial PVA models of arboreal marsupials using significance tests and Bayesian statistics. *Biological Conservation*, **98**, 191–200.

McCullagh, P. and Nelder, J.A. (1989). *Generalised Linear Models*, 2nd edn. New York: Chapman and Hall.

McGarigal, K. and Cushman, S.A. (2002). Comparative evaluation of experimental approaches to the study of fragmentation studies. *Ecological Applications*, **12**, 335–345.

McIntyre, S. (1994). Integrating agriculture and land use and management for conservation of a native grassland flora in a variegated landscape. *Pacific Conservation Biology*, **1**, 236–244.

McIntyre, S. and Barrett, G.W. (1992). Habitat variegation, an alternative to fragmentation. *Conservation Biology*, **6**, 146–147.

McIntyre, S. and Hobbs, R. (1999). A framework for conceptualizing human effects on landscapes and its relevance to management and research models. *Conservation Biology*, **13**, 1282–1292.

McIntyre, S., Barrett, G.W. and Ford, H.A. (1996). Communities and ecosystems. In *Conservation Biology*, ed. I.F. Spellerberg. Harlow, UK: Longman, pp. 154–170.

McIntyre, S., McIvor, J.G. and MacLeod, N.D. (2000). Principles for sustainable grazing in eucalypt woodlands: landscape-scale indicators and the search for thresholds. In *Management for Sustainable Ecosystems*, ed. P. Hale, D. Moloney and P. Sattler. Brisbane, Australia: Centre for Conservation Biology, The University of Queensland, pp. 92–100

McIntyre, S., McIvor, J.G. and Heard, K.M. (eds.) (2002). *Managing and Conserving Grassy Woodlands*. Melbourne, Australia: CSIRO Publishing.

McNeeley, J.A., Neville, L.E. and Rejmanek, M. (2003). When is eradication a sound investment? *Conservation in Practice*, **4**, 30–41.

Means, D.B. and Simberloff, D. (1987). The peninsula effect: habitat-correlated species decline in Florida's herpetofauna. *Journal of Biogeography*, **14**, 551–568.

Menkhorst, P. (ed.) (1995). *Mammals of Victoria: Distribution, Ecology and Conservation*. Melbourne, Australia: Oxford University Press.

Menkhorst, P. (2001). *A Field Guide to the Mammals of Australia*. Melbourne, Australia: Oxford University Press.

Middleton, J. and Merriam, G. (1981). Woodland mice in a farmland mosaic. *Journal of Applied Ecology*, **18**, 703–710.

Millennium Ecosystem Assessment (2005). *Millennium Ecosystem Assessment Synthesis Report* [available at www.millenniumassessment.org].

Mills, L.S. and Allendorf, F.W. (1996). The one-migrant-per-generation rule in conservation and management. *Conservation Biology*, **10**, 1509–1518.

Mirande, C., Lacy, R.C. and Seal, U.S. (1991). *Whooping Crane Population Viability Analysis and Species Survival Plan*. Apple Valley: MN: Captive Breeding Specialist Group, Species Survival Commission, IUCN.

Mönkkönen, M. and Reunanen, P. (1999). On critical thresholds in landscape connectivity: a management perspective. *Oikos*, **84**, 302–305.

Moore, S.E. and Allen, H.L. (1999). Plantation forestry. In *Managing Biodiversity in Forest Ecosystems*, ed. M.L. Hunter. Cambridge, UK: Cambridge University Press, pp. 400–433.

Morgan, G. (2001). *Landscape Health in Australia: A Rapid Assessment of the Relative Condition of Australia's Bioregions and Subregions*. Canberra, Australia: Environment Australia and National Land and Water Resources Audit.

Morrison, M.L., Marcot, B.G. and Mannan, R.W. (2006). *Wildlife–Habitat Relationships: Concepts and Applications*. Washington, DC: Island Press.

Munks, S.A., Mooney, N., Pemberton, D. and Gales, R. (2004). An update on the distribution and status of possums and gliders in Tasmania, including off-shore islands. In *The Biology of Australian Possums and Gliders*, eds. R.L. Goldingay and S.M. Jackson. Sydney, Australia: Surrey Beatty and Sons, pp. 111–129.

Murcia, C. (1995). Edge effects on fragmented forests: implications for conservation. *Trends in Ecology and Evolution*, **10**, 58–62.

Nix, H.A. (1986). A biogeographic analysis of the Australian elapid snakes. In *Atlas of Elapid Snakes*, ed. R. Longmore. Australian Flora and Fauna Series, Number 7. Canberra, Australia: Australian Government Publishing Service, pp. 4–15.

Nix, H.A. and Switzer, M.A. (1991) Rainforest animals: Atlas of vertebrates endemic to Australia's wet tropics. *Kowari*, **1**, 1–112.

Norton, D.A. (1996). Monitoring biodiversity in New Zealand's terrestrial ecosystems. In *Papers from a Seminar Series on Biodiversity*, B. McFadgen and P. Simpson (compilers). Wellington, New Zealand: Department of Conservation, pp. 19–41.

Norton, T.W. (1988). Ecology of greater gliders in different eucalypt forests in south-eastern New South Wales. Unpublished Ph.D. thesis, The Australian National University, Canberra, Australia.

Noske, R.A. (1982). Comparative behaviour and ecology of some Australian bark foraging birds. Unpublished Ph.D. thesis, University of New England, Armidale, NSW, Australia.

Noske, R.A. (1986). Intersexual niche segregation among three bark-foraging birds of eucalypt forests. *Australian Journal of Ecology*, **11**, 255–267.

Noss, R.F. (1987). Corridors in real landscapes: a reply to Simberloff and Cox. *Conservation Biology*, **1**, 159–164.

Noss, R.F. (1991). Landscape connectivity: different functions at different scales. In *Landscape Linkages and Biodiversity*, ed. W.E. Hudson. Covelo, CA: Island Press, pp. 27–39.

O'Grady, J.J., Reed, D.H., Brook, B.W. and Frankham, R. (2004). What are the best correlates of predicted extinction risk? *Biological Conservation*, **118**, 513–520.

Oliver, I. and Beattie, A.J. (1996). Invertebrate morphospecies as surrogates for species: a case study. *Conservation Biology*, **10**, 99–109.

Paetkau, D., Slade, R., Burden, M. and Estoup, A. (2004). Genetic assignment methods for the direct, real-time estimation of migration rate: a simulation-based exploration of accuracy and power. *Molecular Ecology*, **13**, 55–65.

Pahl, L. (1984). Diet preference, diet composition and population density of the ringtail possum (*Pseudocheirus peregrinus cooki*) in several plant communities in southern Victoria. In *Possums and Gliders*, ed. A.P. Smith and I.D. Hume. Sydney, Australia: Surrey Beatty and Sons, pp. 252–260.

Palomares, F., Delibes, M., Ferreras, P., Fedriani, J.M., Calzada, J. and Revilla, E. (2000). Iberian lynx in a fragmented landscape: predispersal, dispersal, and post-dispersal habitats. *Conservation Biology*, **14**, 809–818.

Parker, M. and Mac Nally, R. (2002). Habitat loss and the habitat fragmentation threshold: an experimental evaluation of impacts on richness and total abundances using grassland invertebrates. *Biological Conservation*, **105**, 217–229.

Parr, C.L. and Andersen, A.N. (2006). Patch mosaic burning for biodiversity conservation: a critique of the pyrodiversity paradigm. *Conservation Biology*, **20**, 1610–1619.

Parris, K.M. and Lindenmayer, D.B. (2004). Evidence that creation of a *Pinus radiata* plantation in south-eastern Australia has reduced habitat for frogs. *Acta Oecologia*, **25**, 93–101.

Paton, P.W. (1994). The effect of edge on avian nest success: how strong is the evidence? *Conservation Biology*, **8**, 17–26.

Patterson, B.D. (1987). The principle of nested subsets and its implications for biological conservation. *Conservation Biology*, **1**, 247–293.

Patterson, B.D., and Atmar, W. (1986). Nested subsets and the structure of insular mammalian faunas and archipelagos. *Biological Journal of the Linnean Society*, **28**, 65–82.

Patterson, B.D., and Brown, J.H. (1991). Regionally nested patterns of species composition in granivorous rodent assemblages. *Journal of Biogeography*, **18**, 395–402.

Peakall, R. and Lindenmayer, D.B. (2006). Genetic insights into population recovery following experimental perturbation in a fragmented landscape. *Biological Conservation*, **132**, 520–532.

Peakall, R., Ruibal, M. and Lindenmayer, D.B. (2003). Spatial autocorrelation analysis offers insights into gene flow in the Australian bush rat, *Rattus fuscipes*. *Evolution*, **57**, 1182–1195.

Peakall, R., Ebert, D., Cunningham, R.B. and Lindenmayer, D.B. (2005). Mark–recapture by genetic tagging reveals restricted movements by bush rats, *Rattus fuscipes*, in a fragmented landscape. *Journal of Zoology*, **268**, 207–216.

Peterken, G.F., and Ratcliffe, P.R. (1995). The potential of biodiversity in British upland spruce forests. *Forest Ecology and Management*, **79**, 153–160.

Pharo, E., Lindenmayer, D.B. and Taws, N. (2004). The response of bryophytes to landscape context: a large-scale fragmentation study. *Journal of Applied Ecology*, **41**, 910–921.

Pianka, E.R. (1970). On *r* and *K* selection. *American Naturalist*, **104**, 592–597.

Pither, J. and Taylor, P.D. (1998). An experimental assessment of landscape connectivity. *Oikos*, **83**, 166–174.

Pope, M.L. (2003). A study of the greater glider (*Petauroides volans*) persisting in remnant eucalypt patches surrounded by a softwood plantation matrix. Unpublished M.Sc. thesis, The Australian National University, Canberra, Australia.

Pope, M.L., Lindenmayer, D.B. and Cunningham, R.B. (2004). Patch use by the greater glider (*Petauroides volans*) in a fragmented forest ecosystem. I. Home range size and movements. *Wildlife Research*, **31**, 559–568.

Possingham, H.P. and Davies, I. (1995). ALEX: a model for the viability analysis of spatially structured populations. *Biological Conservation*, **73**, 143–150.

Possingham, H.P., Lindenmayer, D.B. and McCarthy, M.A. 2001. Population viability analysis. In *Encyclopaedia of Biodiversity*, Volume **4**, pp. 831–843.

Press, A.J. (1987). Comparison of the demography of populations of *Rattus fuscipes* living in cool temperate rainforests and dry sclerophyll forests. *Australian Wildlife Research*, **14**, 45–63.

Pressey, R.L. (1995). Conservation reserves in NSW: crown jewels or leftovers? *Search*, **26**, 47–51.

Pressey, R.L., Hagar, T.C., Ryan, K.M., Schwarz, J., Wall, S., Ferrier, S. and Creaser, P.M. (2000). Using abiotic data for conservation assessments over extensive regions: quantitative methods applied across New South Wales, Australia. *Conservation Biology*, **96**, 55–82.

Price, O.F., Woinarski, J.C.Z. and Robinson, D. (1999). Very large requirements for frugivorous birds in monsoon rainforests of the Northern Territory, Australia. *Biological Conservation*, **91**, 169–180.

Pulliam, H.R., Dunning, J.B. and Liu, J. (1992). Population dynamics in complex landscapes: a case study. *Ecological Applications*, **2**, 165–177.

Pullin, A.S. (2002). *Conservation Biology*. Cambridge, UK: Cambridge University Press.

Pyke, G.H. and Recher, H.F. (1983). Censusing Australian birds: a summary of procedures and a scheme for standardisation of data presentation and storage. In *Methods of Censusing Birds in Australia*, ed. S.J. Davies. Proceedings of a Symposium organised by the Zoology section of the ANZAAS and the Western Australian Group of the Royal Australasian Ornithologists Union. Perth, Australia: Department of Conservation and Environment, pp. 55–63.

Radford, J.Q., Bennett, A.F. and Cheers, G.J. (2005). Landscape-level thresholds of habitat cover for woodland-dependent birds. *Biological Conservation*, **124**, 317–337.

Raivio, S. (1988). The peninsular effect and habitat structure: bird communities in coniferous forests of the Hanko peninsula, southern Finland. *Ornis Fennica*, **65**, 129–149.

Recher, H.F. and Holmes, R.T. (2000). The foraging ecology of birds of eucalypt forest and woodland. I. Differences between males and females. *Emu*, **100**, 205–215.

Recher, H.F., Rohan-Jones, W. and Smith, P. (1980). *Effects of the Eden Woodchip Industry on Terrestrial Vertebrates, with Recommendations for Management*. Research Note No. 42. Sydney, Australia: Forest Commission of New South Wales.

Recher, H.F., Holmes, R.T., Schulz, M., Shields, J. and Kavanagh, R. (1985). Foraging patterns of breeding birds in eucalypt forest and woodland of southeastern Australia. *Australian Journal of Ecology*, **10**, 399–419.

Recher, H.F., Shields, J., Kavanagh, R.P. and Webb, G. (1987). Retaining remnant mature forest for nature conservation at Eden, New South Wales: a review of theory and practice. In *Nature Conservation: The Role of Remnants of Vegetation*, ed. D.A. Saunders, G.W. Arnold, A.A. Burbidge and A.J. Hopkins. Chipping Norton, NSW, Australia: Surrey Beatty and Sons, pp. 177–194.

Reed, D.H., O'Grady, J.J., Brook, B.W., Ballou, J.D. and Frankham, R. (2003). Estimates of minimum viable population sizes for vertebrates and factors influencing those estimates. *Biological Conservation*, **113**, 23–34.

Reed, J.M. and Dobson, A.P. (1993). Behavioural constraints and conservation biology: conspecific attraction and recruitment. *Trends in Evolution and Ecology*, **8**, 253–256.

Reid, J.R. (1999). *Threatened and Declining Birds in the New South Wales Sheep–Wheat Belt. I. Diagnosis, Characteristics and Management*. Consulting Report. Canberra, Australia: NSW National Parks and Wildlife Service.

Renjifo, L.M. (2001). Effect of natural and anthropogenic landscape matrices on the abundance of sub-Andean bird species. *Ecological Applications*, **11**, 14–31.

Richardson, D.M., Williams, P.A. and Hobbs, R.J. (1994). Pine invasions in the Southern Hemisphere: determinants of spread and invadability. *Journal of Biogeography*, **21**, 511–527.

Ricketts, T.H. (2001). The matrix matters: effective isolation in fragmented landscapes. *American Naturalist*, **158**, 87–99.

Ries, L., Fletcher, R.J., Battin, J. and Sisk, T.D. (2004). Ecological responses to habitat edges: mechanisms, models, and variability explained. *Annual Review of Ecology Evolution and Systematics*, **35**, 491–522.

Rishworth, C., McIlroy, J. and Tanton, M.T. (1995a). Diet of the common wombat, *Vombatus ursinus*, in plantations of *Pinus radiata*. *Wildlife Research*, **22**, 333–339.

Rishworth, C., McIlroy, J. and Tanton, M.T. (1995b). Factors effecting population densities of the common wombat, *Vombatus ursinus*, in plantations of *Pinus radiata*. *Forest Ecology and Management*, **76**, 11–19.

Roberge, J.-M. and Angelstam, P. (2004). Usefulness of the umbrella species concept as a conservation tool. *Conservation Biology*, **18**, 76–85.

Robinson, A.C. (1987). The ecology of the bush rat, *Rattus fuscipes* (Rodentia: Muridae), in Sherbrooke Forest, Victoria. *Australian Mammalogy*, **11**, 35–49.

Robinson, G.R., Holt, R.D., Gaines, M.S., Hamburg, S.P., Johnson, M.L., Fitch, H.S. and Martinko, E.A. (1992). Diverse and contrasting effects of habitat fragmentation. *Science*, **257**, 524–526.

Rolstad, J. and Wegge, P. (1987). Distribution and size of capercaillie leks in relation to old forest fragmentation. *Oecologia*, **72**, 389–394.

Rosenblatt, D.L., Heske, E.J., Nelson, S.L., Barber, D.H., Miller, M.A. and MacAllister, B. (1999). Forest fragments in east-central Illinois: islands or habitat patches for mammals? *American Midland Naturalist*, **141**, 115–123.

Rosenzweig, M.L. (1995). *Species Diversity in Space and Time*. Cambridge UK: Cambridge University Press.

Rübsamen, K., Hume, I.D., Foley, W.J. and Rübsamen, U. (1984). Implications of the large surface area to body mass ratio on the heat balance of the greater glider (*Petauroides volans*: Marsupialia). *Journal of Comparative Physiology*, **154**, 105–111.

Saccheri, I., Kuussaari, M., Kankare, M., Vikman, P., Fortelius, W. and Hanski, I. (1998). Inbreeding and extinction in a butterfly metapopulation. *Nature*, **392**, 491–494.

Sala, O.E., Chapin, F.S., Armesto, J.J., Berlow, E., Bloomfield, J., Dirzo, R., Huber-Sanwald, E., Huenneke, L.F., Jackson, R.B., Kinzig, A., Leemans, R., Lodge, D. M., Mooney, H.A., Oesterheld, M., Poff, N.L., Sykes, M.T., Walker, B.H., Walker, M. and Wall, D.H. (2000). Biodiversity: Global biodiversity scenarios for the year 2100. *Science*, **287**, 1770–1774.

Salt, D., Lindenmayer, D.B. and Hobbs, R.J. (2004). *Trees and Biodiversity: A Guide for Australian Farm Forestry*. Canberra, Australia: Rural Industries Research and Development Corporation.

Sanecki, G.M., Green, K., Wood, H., Lindenmayer, D.B. and Sanecki, K.L. (2006). The influence of snow cover on home range and activity of the bush rat (*Rattus fuscipes*) and the dusky antechinus (*Antechinus swainsonii*). *Wildlife Research*, **33**, 489–496.

Sargent, R.A., Kilgo, J.C., Chapman, B.R. and Miller, K.V. (1998). Predation of artificial nests in hardwood fragments enclosed by pine and agricultural habitats. *Journal of Wildlife Management*, **62**, 1438–1442.

Sarre, S., Smith, G.T. and Meyers, J.A. (1995). Persistence of two species of gecko (*Oedura reticulata* and *Gehyra variegata*) in remnant habitat. *Biological Conservation*, **71**, 25–33.

Sasaki, T., Okayasu, T., Jamsran, U. and Takeuchi, K. (2008). Threshold changes in vegetation along a grazing gradient in Mongolian rangelands. *Journal of Ecology*, **96**, 145–154.

Saunders, D.A., Arnold, G.W., Burbidge, A.A. and Hopkins, A.J. (eds.) (1987). *Nature Conservation: The Role of Remnants of Native Vegetation*. Chipping Norton, NSW, Australia: Surrey Beatty and Sons.

Saunders, D.A., Hobbs, R.J. and Margules, C.R. (1991). Biological consequences of ecosystem fragmentation: a review. *Conservation Biology*, **5**, 18–32.

Savolainen, P., Leitner, T., Wilton, A.N., Matisoo-Smith, E. and Lundeberg, J. (2004). A detailed picture of the origin of the Australian dingo, obtained from the study of mitochondrial DNA. *Proceedings of the National Academy of Sciences*, **101**, 12387–12390.

Scheffer, M., Carpenter, S., Foley, J.A., Folke, C. and Walker, B. (2001). Catastrophic shifts in ecosystems. *Nature*, **413**, 591–596.

Scheiner, S.M. (2003). Six types of species–area curves. *Global Ecology and Biogeography*, **12**, 441–447.

Schilthuizen, M., Liew, T.-S., Elahan, B.B. and Lackman-Ancrenaz, I. (2005). Effects of karst forest degradation on pulmonate and prosobranch land snail communities in Sabah, Malaysian Borneo. *Conservation Biology*, **19**, 949–954.

Schmuki, C., Vorburger, C., Runciman, D., MacEachern, S. and Sunnucks, P. (2006a). When log-dwellers meet loggers: impacts of forest fragmentation on two endemic log-dwelling beetles in southeastern Australia. *Molecular Ecology*, **15**, 1481–1492.

Schmuki, C., Blacket, M.J. and Sunnucks, P. (2006b). Anonymous single-copy nuclear DNA (scnDNA) markers for two endemic log-dwelling beetles: *Apasis puncticeps* and *Adelium calosomoides* (Tenebrionidae: Lagriinae: Adeliini). *Molecular Ecology Notes*, **6**, 362–364.

Schoener, T.W. and Schoener, A. (1983). Distribution of vertebrates on some very small islands. I. Occurrence sequences of individual species. *Journal of Animal Ecology*, **52**, 209–235.

Scott, J.M., Abbitt, R.J. and Groves, C.R. (2001). What are we protecting? The United States conservation portfolio. *Conservation Biology in Practice*, 2, 18–19.

Scott, L.K., Hume, I.D. and Dickman, C.R. (1999). Ecology and population biology of long-nosed bandicoots (*Perameles nasuta*) at North Head, Sydney Harbour National Park. *Wildlife Research*, 26, 805–821.

Seebeck, J.H., Warneke, R.M. and Baxter, B.J. (1984). Diet of the bobuck, *Trichosurus caninus* (Ogilby) (Marsupialia: Phalangeridae) in a mountain forest in Victoria. In *Possums and Gliders*, ed. A.P. Smith and I.D. Hume. Sydney, Australia: Surrey Beatty and Sons, pp. 145–154.

Seib, R.L. (1980). Baja California: a peninsula for rodents but not for reptiles. *American Naturalist* 115, 613–620.

Shafer, C.L. (1990). *Nature Reserves: Island Theory and Conservation Practice*. Washington, DC: Smithsonian Institution Press.

Shakesby, R.A., Boakes, D.J., Coelho, C. de O.A., Bento Gonçalves, A.J. and Walsh, R.P.D. (1996). Limiting the soil degradational impacts of wildfire in pine and eucalyptus forests in Portugal: a comparison of alternative post-fire management practices. *Applied Geography*, 16, 337–355.

Shiel, D. and Burslem, F.R. 2003. Disturbing hypotheses in tropical forests. *Trends in Ecology and Evolution*, 18, 18–26.

Siitonen, P., Lehtinen, A. and Siitonen, M. (2005). Effects of forest edges on the distribution, abundance and regional persistence of wood-rotting fungi. *Conservation Biology*, 19, 250–260.

Simberloff, D.A. (1988). The contribution of population and community biology to conservation science. *Annual Review of Ecology and Systematics*, 19, 473–511.

Simberloff, D.A. (1998). Flagships, umbrellas, and keystones: is single-species management passé in the landscape era. *Biological Conservation*, 83, 247–257.

Simberloff, D. and Abele, L.G. (1982). Refuge design and island geographic theory: effects of fragmentation. *American Naturalist*, 120, 41–45.

Simberloff, D. and Levin, B. (1985). Predictable sequences of species loss with decreasing island area: land birds in two archipelagoes. *New Zealand Journal of Ecology*, 8, 11–20.

Simberloff, D. and Martin, J.L. (1991). Nestedness of insular avifaunas: simple summary statistics masking species patterns. *Ornis Fennica*, 68, 178–192.

Simberloff, D.A., Farr, J.A., Cox, J. and Mehlman, D.W. (1992). Movement corridors: conservation bargains or poor investments? *Conservation Biology*, 6, 493–504.

Simpson, G.G. (1964). Species density of North American recent mammals. *Systematic Zoology*, 13, 57–73.

Sisk, T.D., Noon, B.R. and Hampton, H.M. (2002). Estimating the effective area of habitat patches in heterogeneous landscapes. In *Predicting Species Occurrences: Issues of Accuracy and Scale*, ed. M. Scott, P. Heglund, M.L. Morrison, J.B. Haufler, M.G. Raphael, W.A. Wall and F.B. Samson. Washington, DC: Island Press, pp. 713–725.

Sjorgen-Gulve, P. (1994). Distribution and extinction patterns within a Northern metapopulation of the pool frog, *Rana lessonae*. *Ecology*, 75, 1357–1367.

Smith, A.T. and Peacock, M.M (1990). Conspecific attraction and the determination of metapopulation colonization rates. *Conservation Biology*, 4, 320–323.

Smith, B., Augee, M. and Rose, S. (2003). Radio-tracking studies of common ringtail possums, *Pseudocheirus peregrinus*, in Manly Dam Reserve, Sydney. *Proceedings of the Linnean Society of New South Wales*, **124**, 183–194.

Smith, P.A. (1994). Autocorrelation in logistic regression modeling of species' distribution. *Global Ecology and Biogeography Letters*, **4**, 47–61.

Smith, P.L. (2006). Habitat islands in real landscapes, the influence of the matrix. Unpublished Ph.D. thesis, Macquarie University, Sydney, Australia.

Smith, R.F. (1969). Studies of the marsupial glider, *Schoinobates volans*. I. Reproduction. *Australian Journal of Zoology*, **17**, 625–636.

Smith, R.J. (2000). An Investigation into the relationships between anthropogenic forest disturbance patterns, population viability and landscape indices. Unpublished M.Sc. thesis, Institute of Land and Food Resources, The University of Melbourne, Australia.

Smyth, A.K., Lamb, D., Hall, L., McCallum, H., Moloney, D. and Smith, G. 2000. Towards scientifically valid management tools for sustainable forest management: species guilds versus model species. In *Management for Sustainable Ecosystems*, ed. P. Hale, A. Petrie, D. Moloney and P. Sattler. Brisbane, Australia: Centre for Conservation Biology, The University of Brisbane, pp. 118–124.

Soderquist, T.R. and Mac Nally, R. (2000). The conservation value of mesic gullies in dry forest landscapes: mammal populations in the box–ironbark ecosystem of southern Australia. *Biological Conservation*, **93**, 281–291.

Starfield, A.M. and Bleloch, A.L. (1992). *Building Models for Conservation and Wildlife Management*. Edina, MN: Burgess International Group.

Stenseth, N. and Lidicker, W. (eds.) (1992a). *Animal Dispersal*. London: Chapman and Hall.

Stenseth, N. and Lidicker, W. (eds.) (1992b). Appendix 1: Where do we stand methodologically about experimental design and methods of analysis in the study of dispersal? In *Animal Dispersal*, ed. N. Stenseth and W. Lidicker. London: Chapman and Hall, pp. 295–315.

Stouffer, P.C., Bierregaard, R.O., Strong, C. and Lovejoy, T. (2006). Long-term landscape change and bird abundance in Amazonian rainforest fragments. *Conservation Biology*, **20**, 1212–1223.

Strahan, R. (1995). *Complete Book of Australian Mammals*. Sydney, Australia: Angus and Robertson.

Strahler, A.N. (1957). Quantitative analysis of watershed geomorphology. *Transactions of the American Geophysical Union*, **38**, 913–920.

Suckling, G.C. (1982). Value of reserved habitat for mammal conservation in plantations. *Australian Forestry*, **45**, 19–27.

Suckling, G.C., Backen, E., Heislers, A. and Neumann, F.G. (1976). *The Flora and Fauna of Radiata Pine Plantations in North-eastern Victoria*. Forest Commission of Victoria Bulletin No. 24. Melbourne, Australia: Forest Commission of Victoria.

Tang, S.M. and Gustafson, E.J. (1997). Perception of scale in forest management planning: challenges and implications. *Landscape and Urban Planning*, **39**, 1–9.

Taylor, A.C., Alpers, D. and Sherwin, W.B. (1998). Remote censusing of northern hairy-nosed wombats *Lasiorhinus krefftii* via genetic typing of hairs collected in the field. In *Wombats*, ed. R.T. Wells and P.A. Pridmore. Chipping Norton, NSW, Australia: Surrey Beatty and Sons, pp. 156–164.

Taylor, A.C., Tyndale-Biscoe, H. and Lindenmayer, D.B. (2007). Unexpected persistence on habitat islands: genetic signatures reveal dispersal of a eucalypt-dependent marsupial through a hostile pine matrix. *Molecular Ecology*, **16**, 2655–2666.

Taylor, B.L. (1995). The reliability of using population viability analysis for risk classification of species. *Conservation Biology*, **9**, 551–558.

Taylor, M. and Horner, B.E. (1973). Reproductive characteristics of wild native Australian *Rattus* (Rodentia: Muridae). *Australian Journal of Zoology*, **21**, 437–475.

Taylor, R.J. and Regal, P.J (1978). The peninsula effect on species diversity and the biogeography of Baja California. *American Naturalist*, **112**, 583–593.

Temple, S.A. and Cary, J.R. (1988). Modelling dynamics of habitat interior bird populations in fragmented landscapes. *Conservation Biology*, **2**, 340–347.

Terborgh, J. (1989). *Where Have all the Birds Gone?* Princeton, NJ: Princeton University Press.

Thomson, J.A. and Owen, W.H (1964). A field study of the Australian ringtail possum *Pseudocheirus peregrinus* (Marsupialia: Phalangeridae). *Ecological Monographs*, **34**, 27–52.

Tickle, P., Hafner, S., Lesslie, R., Lindenmayer, D.B., McAlpine, C., Mackey, B., Norman, P. and Phinn, S. (1998). *Scoping Study: Montreal Indicator 1.1e. Fragmentation of Forest Types – Identification of Research Priorities*. Final Report. Canberra, Australia: Forest and Wood Products Research and Development Corporation.

Tilman, D., May, R.M., Lehman, C.L. and Nowak, M.A. (1994). Habitat destruction and the extinction debt. *Nature*, **371**, 65–66.

Tischendorf, L. and Fahrig, L. (2000a). On the usage and measurement of landscape connectivity. *Oikos*, **90**, 7–19.

Tischendorf, L. and Fahrig, L. (2000b). How should we measure landscape connectivity? *Landscape Ecology*, **15**, 235–254.

Tubelis, D.P., Lindenmayer, D.B., Saunders, D.A., Cowling, A. and Nix, H.A. (2004). Landscape supplementation provided by an exotic matrix: implications for bird conservation and forest management in a softwood plantation system in south-eastern Australia *Oikos*, **107**, 634–644.

Tubelis, D.P., Lindenmayer, D.B. and Cowling, A. (2007a). Bird populations in native forest patches in south-eastern Australia: the roles of patch width, matrix type (age) and matrix use. *Landscape Ecology*, **22**, 1045–1058.

Tubelis, D.P., Lindenmayer, D.B. and Cowling, A. (2007b). The peninsula effect on bird species in native eucalypt forests in an Australian wood production landscape. *Journal of Zoology*, **271**, 11–18.

Turner, I.M. (1996). Species loss in fragments of tropical rain forest: a review of the evidence. *Journal of Applied Ecology*, **33**, 200–209.

Turner, M.G. (1989). Landscape ecology: the effect of pattern on process. *Annual Review of Ecology and Systematics*, **20**, 171–197.

Turner, M.G., Romme, W.H. and Tinker, D.B. (2003). Surprises and lessons from the 1988 Yellowstone fires. *Frontiers in Ecology and the Environment*, **1**, 351–358.

Tyler, M.J. (1999). *Australian Frogs: A Natural History*. Sydney, Australia: Reed New Holland.

Tyndale-Biscoe, C.H. and Smith, R.F.C. (1969a). Studies on the marsupial glider *Schoinobates volans* (Kerr). II. Population structure and regulatory mechanisms. *Journal of Animal Ecology*, **38**, 637–650.

Tyndale-Biscoe, C.H. and Smith, R.F.C. (1969b) Studies on the marsupial glider *Schoinobates volans* (Kerr). III. Response to habitat destruction. *Journal of Animal Ecology*, **38**, 651–659.

United Nations Environment Program (UNEP) (1999). *Global Environmental Outlook 2000*. Nairobi, Kenya: United Nations Environment Program.

Viggers, K.L. and Lindenmayer, D.B. (2000). A population study of the mountain brushtail possum (*Trichosurus caninus*) in the central highlands of Victoria. *Australian Journal of Zoology*, **48**, 201–216.

Viggers, K.L. and Lindenmayer, D.B. (2001). Haematological and plasma biochemical values for the greater glider (*Petauroides volans*). *Journal of Wildlife Diseases*, **37**, 370–374.

Viggers, K.L. and Lindenmayer, D.B. (2004). A review of the biology of the short-eared possum *Trichosurus caninus* and the mountain brushtail possum *Trichosurus cunninghamii*. In *Possums and Gliders*, ed. R.L. Goldingay and S. Jackson. Sydney, Australia: Surrey Beatty and Sons, pp. 490–505.

Villard, M.A. (1998). On forest-interior species, edge avoidance, area sensitivity, and dogmas in avian conservation. *Auk*, **115**, 801–805.

Villard, M.A. and Taylor, P.D. (1994). Tolerance to habitat fragmentation influences the colonization of new habitat by forest birds. *Oecologia*, **98**, 393–401.

Villard, M.A., Trzcinski, M.K. and Merriam, G. (1999). Fragmentation effects on forest birds: relative influence of woodland cover and configuration on landscape occupancy. *Conservation Biology*, **13**, 774–783.

Walker, B. and Meyers, J.A. (2004). Thresholds in ecological and social–ecological systems: a developing database. *Ecology and Society*, **9**, 3 [available at www.ecologyandsociety.org/vol9/iss2/art3].

Walters, C.J. 1986. *Adaptive Management of Renewable Resources*. Macmillan Publishing Company, New York.

Warneke, R.M. (1971). Field study of the bush rat (*Rattus fuscipes*). *Wildlife Contributions Victoria*, **14**, 1–115.

Watson, J., Freudenberger, D. and Paull, D. (2001). An assessment of the focal-species approach for conserving birds in variegated landscapes in southeastern Australia. *Conservation Biology*, **15**, 1364–1373.

Watson, J.E., Whittaker, R.J. and Freudenberger, D. (2005). Bird community responses to habitat fragmentation: how consistent are they across landscapes? *Journal of Biogeography*, **32**, 1353–1370.

Watts, C.H. and Aslin, H. (1981). *The Rodents of Australia*. Hong Kong: Angus and Robertson.

Wayne, A.F. (2005). The ecology of the Koomal (*Trichosurus vulpecula hypoleucus*) and Ngwayir (*Pseudocheirus occidentalis*) in the Jarrah forests of south-western Australia. Unpublished Ph.D. thesis, Centre for Resource and Environmental Studies, The Australian National University, Canberra Australia.

Wegner, J. (1994). *Ecological Landscape Variables for Monitoring and Management of Forest Biodiversity in Canada*. Report to Canadian Ministry of Natural Resources. Manotick, Canada: GM Group, Ecological Land Management.

Wethered, R. and Lawes, M.J. (2003). Matrix effects on bird assemblages in fragmented afromontane forests in South Africa. *Conservation Biology*, **114**, 327–340.

Whittaker, R.J. (1998). *Island Biogeography: Ecology, Evolution and Conservation.* Oxford, UK: Oxford University Press.

Wiens, J. (1994). Habitat fragmentation: island vs landscape perspectives on bird conservation. *Ibis,* **137,** S97–S104.

Wiens, J. (1995). Landscape mosaics and ecological theory. In *Landscape Mosaics and Ecological Processes,* ed. L. Hansson, L. Fahrig and G. Merriam. London: Chapman and Hall, pp. 1–26.

Wiens, J. (1997). Wildlife in patchy environments: metapopulations, mosaics and management. In *Metapopulations and Wildlife Conservation,* ed. D.R. McCullogh. Covelo, CA: Island Press, pp. 53–84.

Wiens, J. (1999). The science and practice of landscape ecology. In *Landscape Ecological Analysis: Issues and Applications,* ed. J.M. Klopatek and R.H. Gardner. New York: Springer, pp. 37–383.

Wiens, J.A., Schooley, R.L. and Weekes, R.D. (1997). Patchy landscapes and animal movements: do beetles percolate? *Oikos,* **78,** 257–264.

Wiggins, D.A. (1999). The peninsula effect on species diversity: a reassessment of the avifauna of Baja California. *Ecography* **22,** 542–547.

Wilcove, D.S. (1985). Nest predation in forest tracts and the decline of migratory songbirds. *Ecology,* **66,** 1211–1214.

Williams, S.E. and Hero, J.M. (2001). Multiple determinants of Australian tropical frog biodiversity. *Biological Conservation,* **98,** 1–10.

Wilson, S. and Swan, G. (2003). *A Complete Guide to the Reptiles of Australia.* Sydney, Australia: Reed New Holland.

With, K.A. and Crist, T.O. (1995). Critical thresholds in species' responses to landscape structure. *Ecology,* **76,** 2446–2459.

With, K.A. and King, A.W. (1999). Extinction thresholds for species in fractal landscapes. *Conservation Biology,* **13,** 314–326.

Witham, T.G., Morrow, P.A. and Potts, B.M. (1991). Conservation of hybrid plants. *Science,* **254,** 779–780.

Wolff, J.O., Schauber, E.M. and Edge, W.D. (1997). Effects of habitat loss and fragmentation in the behavior and demography of gray-tailed voles. *Conservation Biology,* **11,** 945–956.

Worthen, W.B. (1996). Community composition and nested-subset analyses: basic descriptors for community ecology. *Oikos,* **76,** 417–426.

Wright, D.H., Patterson, B.D., Mikkelson, G.M. Cutler, A. and Atmar, W. (1998). A comparative analysis of nested subset patterns of species composition. *Oecologia,* **113,** 1–20.

Wu, J. and Loucks, O.L. (1995). From balance of nature to hierarchical patch dynamics: a paradigm shift in ecology. *Quarterly Review of Biology,* **70,** 439–466.

Yamaura, Y., Katoh, K. and T. Takahashi, T. (2006). Reversing habitat loss: deciduous habitat fragmentation matters to birds in a larch plantation matrix. *Ecography,* **29,** 827–834.

Youngentob, K., Lindenmayer, D.B., Held, A.A. and Jia, X. (2008). A prospective study of the effects of foliage chemistry and landscape context on the distribution and abundance of arboreal marsupials near Tumut, New South Wales. In *The 13th Australasian Remote Sensing and Photogrammetry Conference,* ed. W. Cartwright, G. Gartner, L. Meng, and M.P. Peterson. Springer Verlag Series in Geoinformation and Cartography. Heidelberg, Germany: Springer

Zanette, L. and Jenkins, B. (2000). Nesting success and nest predators in forest fragments: a study using real and artificial nests. *Auk*, **117**, 445–454.

Zanette, L., Doyle, P. and Tremont, S.M. (2000). Food shortage in small fragments: evidence from an area-sensitive passerine. *Ecology*, **81**, 1654–1666.

Zanette, L., MacDougall-Shakleton, E., Clinchy, M. and Smith, J.N. (2005). Brown-headed cowbirds skew host offspring sex ratios. *Ecology*, **86**, 815–820.

Zanuncio, J.C., Mezzomo, J.A., Guedes, R.N.C. and Oliveira, A.C. (1998). Influence of strips of native vegetation on Lepidoptera associated with *Eucalyptus cloeziana* in Brazil. *Forest Ecology and Management*, **108**, 85–90.

Zimmerman, B.L. and Bierregaard, R.O. (1986). Relevance of equilibrium theory of island biogeography and species–area relations to conservation with a case study from Amazonia. *Journal of Biogeography*, **13**, 133–143.

Zuidema, P.A., Sayer, J. and Dijkman, W. (1996). Forest fragmentation and biodiversity: the case for intermediate-sized reserves. *Environmental Conservation*, **2**, 290–9

Index

Printed in the United States
By Bookmasters